Philosophy of Engineering and Technology

Volume 49

The *Philosophy of Engineering and Technology* book series provides the multifaceted and rapidly growing discipline of the philosophy of technology with a central overarching and integrative platform. It publishes on all topics in the philosophy of technology and is open to all research communities across the world.

Specifically, it publishes edited volumes and monographs in:

- the phenomenology, anthropology and socio-politics of technology and engineering
- the emergent fields of the ontology and epistemology of technology and design
- engineering ethics and the ethics of specific technologies ranging from nuclear technologies to artificial intelligence
- written from philosophical and practitioners' perspectives and authored by philosophers and practitioners

The series also welcomes proposals that bring these fields together or advance philosophy of engineering and technology in other integrative ways.

Proposals should include:

- A short synopsis of the work or the introduction chapter
- The proposed Table of Contents
- The CV of the lead author(s) or editor(s)
- If available: one sample chapter

We aim to make a first decision within 1 month of submission. In case of a positive first decision the work will be provisionally contracted; the final decision about publication will depend upon the result of anonymous peer review of the complete manuscript. We aim to have the complete work peer-reviewed within 3 months of submission.

The series discourages the submission of manuscripts that contain reprints of previous published material and/or manuscripts that are below 150 pages/75,000 words.

For inquiries and submission of proposals authors can contact the editor-in-chief Pieter Vermaas via: p.e.vermaas@tudelft.nl, or contact one of the associate editors.

Anders Buch • Steen Hyldgaard Christensen
Editors

Bildung for Engineering Education and Practice

A New Agenda

Springer

Editors
Anders Buch
VIA University College
Aarhus, Denmark

Steen Hyldgaard Christensen
Independent researcher
Herning, Denmark

Editor-in-Chief
Pieter E. Vermaas

ISSN 1879-7202 ISSN 1879-7210 (electronic)
Philosophy of Engineering and Technology
ISBN 978-3-031-86580-0 ISBN 978-3-031-86581-7 (eBook)
https://doi.org/10.1007/978-3-031-86581-7

This Springer imprint is published by the registered company Springer Nature Switzerland AG
The registered company address is: Gewerbestrasse 11, 6330 Cham, Switzerland

From the Editor in Chief

The posthumous publication of Dr. Steen Hyldgaard Christensen's sixth volume in the book series *Philosophy of Engineering and Technology* calls for sincerely sadness and deep respect. Steen Hyldgaard Christensen should have been able to witness the publication of this volume since it is completing his decades-long effort to build up an academic corpus and community for the philosophy of engineering and education. Steen Hyldgaard Christensen has been impressively successful with this effort. He created a large, vibrant, and global research network in philosophy of engineering. With the current volume "*Bildung* for Engineering Education and Practice: A New Agenda," with five other volumes in the book series *Philosophy of Engineering and Technology*, and with his publications in other venues, he leaves behind more than 200 papers on the topic. That body of work will fuel research in his field for the decades to come.

I want to thank Anders Buch, co-editor of this volume, for finalizing work on the volume. And I want to express my respect and gratitude for the lasting contribution Dr. Steen Hyldgaard Christensen has made to establishing the field of philosophy of engineering.

Delft, The Netherlands

Pieter Vermaas

Preface

The inspiration for this anthology stems from the recognition that the German concept of *Bildung* has emerged as a significant threshold concept in science and environmental studies. In these fields critical thinking, societal values, ethics, and social responsibility have been introduced as new knowledge domains. However, the historical semantics of the concept are complex. A closer examination reveals that, on the one hand, *Bildung* has served a marker of crisis in educational systems, standing in contrast to contemporary educational practices focused on STEMM (Science, Technology, Engineering, Mathematics & Medicine) fields driven by market imperatives and the neoliberal management of human capital. On the other hand, the chapters in this volume each explore whether reforming engineering education and practice using the concept of *Bildung* as a framework would be advantageous in addressing socially pertinent challenges on a global scale, such as global warming, decreasing biodiversity, and increasing social and economic polarization. All contributors, each in their own way, respond affirmatively.

With backgrounds in engineering, social sciences, and the humanities, we have been developing and/or teaching courses for engineering students across the liberal arts-engineering divide for decades. Through participation in research projects aimed at understanding and reforming engineering education and practice we have published a series of books on engineering education and practice with contributions from a diverse network of scholars globally. This present anthology aims to consolidate the various strands of this collaborative trajectory and to expand it by reflecting on the prospects of reforming engineering education and practice through the new agenda proposed by the title of this volume. Previous publications leading up to this volume include:

- Steen Hyldgaard Christensen, Bernard Delahousse, and Martin Meganck (Eds.) (2007). *Philosophy in Engineering*. Aarhus: Academica.
- Steen Hyldgaard Christensen, Bernard Delahousse, and Martin Meganck (Eds.) (2009). *Engineering in Context.* Aarhus: Academica.

- Steen Hyldgaard Christensen, Carl Mitcham, Li Bocong, and Yanming An (Eds.) (2012). *Engineering, Development and Philosophy: American, Chinese, and European Perspectives*, Cham: Springer.
- Steen Hyldgaard Christensen, Christelle Didier, Andrew Jamison, Martin Meganck, Carl Mitcham, and Byron Newberry (Eds.) (2015). *International Perspectives on Engineering Education; Engineering Education and Practice in Context, Volume I,* Cham: Springer.
- Steen Hyldgaard Christensen, Christelle Didier, Andrew Jamison, Martin Meganck, Carl Mitcham, and Byron Newberry (Eds.) (2015). *Engineering Identities, Epistemologies and Values*: *Engineering Education and Practice in Context. Volume II,* Cham: Springer.
- Steen Hyldgaard Christensen, Bernard Delahousse, Christelle Didier, Martin Meganck, and Mike Murphy (Eds.) (2019). *The Engineering-Business Nexus: Symbiosis, Tension and Co-Evolution.* Cham: Springer.
- Steen Hyldgaard Christensen, Anders Buch, Eddie Conlon, Christelle Didier, Carl Mitcham, and Mike Murphy (Eds.) (2022). *Engineering, Social Sciences, and the Humanities: Have Their Conversations Come of Age?* Cham: Springer.

It has been a priority for us to ensure that the team of scholars and researchers contributing to the anthology is composed of individuals from diverse cultural backgrounds, with close to equal representation of female and male researchers and balanced across generations. The anthology is addressed to both teachers and students in engineering and non-engineering disciplines, as well as to practitioners and educational policymakers at political and institutional levels. It is the result of an extensive writing and editorial process. We hope readers will find it valuable, especially as it aims to inspire us all to engage in deeper thinking and reflection on broadening engineering education and practice through interdisciplinary conversations among engineering, social sciences, and the humanities and to catalyze further research in this important interdisciplinary field.

Herning, Denmark — Steen Hyldgaard Christensen
Aarhus, Denmark — Anders Buch
April 2024

In Memoriam: Dr. Steen Hyldgaard Christensen (1950–2024)

The unexpected death of Steen came as a shock to all who knew him. After a short period of illness, he passed away in late June 2024. During the last 20 years, he had become a well-known and cherished figure among scholars with a philosophical interest in engineering. Steen's academic ambition was profound, consistent, and generous. In 2012, he articulated it himself:

> The aim that I have been pursuing since I started my research on engineering and engineering education in 2005 has been to further the dialogue between engineering and philosophy – philosophy broadly construed. A main concern for me has been to explore ways the humanities can contribute to self-development in engineering education, partly through the appreciation of the multiple contexts within which engineers increasingly work, and partly through the appreciation of the challenges with which engineers, engineering institutions, and engineering education are currently confronted. In so doing I have been striving through a number of initiatives to create conversation spaces in terms of both national and international workshops and research projects where this dialogue between engineers, humanists, and social scientists could take place in a meaningful way. As a result, I have been able to build and sustain a still expanding community of engineering education researchers across the liberal arts-engineering divide. (Christensen, 2012, 1)

There can be no doubt that Steen fulfilled his ambition. Over two decades, he managed to edit and co-author nine international academic anthologies (Christensen et al., 2003, 2007, 2009, 2012, 2015a, 2015b, 2019, 2022, and Buch & Christensen 2025), co-author a historical introduction to science and technology studies (Jamison et al., 2011), initiate and manage numerous national and international research projects, and publish several scientific journal articles, mostly co-authored with other scholars in the field. He made this community-building endeavor the central theme of his PhD thesis, which he submitted to Aalborg University at the age of 62. These achievements were driven by his "hunger" for scholarly work and his aspiration to bring people together to create an intellectual environment that would satisfy his longing for stimulating conversations and critical thinking.

Coming from a non-academic family background, Steen nevertheless started his career with a university degree in literature and the history of ideas. However, job opportunities were scarce, and he had to begin his career in positions that did not reflect his intellectual interests. His career as a teacher and scholar only started in

1987 when he began teaching literature and the history of ideas in an admission program for prospective engineering students at a vocational engineering school in Denmark. Here, he was struck by the cultural clashes between engineering and the humanities and the institutional conflicts that tended to erupt when social and cultural issues were introduced into the engineering curriculum. As the school was later upgraded to an institute of business and technology, with research efforts expected from the staff, Steen seized the opportunity to initiate a curriculum development project that allowed him to explore the professional culture of engineering in greater depth. When the school eventually became part of Aarhus University, Steen could finally engage in the academic research community he had longed to join.

Still, the intellectual environment of his department did not satisfy his intellectual hunger, and he began to reach out and build an academic community that could accommodate his ambitions and intellectual longings. Through extensive reading, he acquainted himself with the literature that tried to bridge the gap between the two cultures and examined the historical and cultural mechanisms that perpetuate this divide. As manager of research projects and editor of anthologies, he reached out to scholars in Europe, America, China, and Australia to broaden the intellectual conversation and include new perspectives, ideas, and voices.

In many ways, Steen pursued his career in an institutional environment that did not support his relentless curiosity and aspirations to transgress established disciplinary boundaries. Nevertheless, his energy and enthusiasm did not wane despite the hostilities he encountered. After retiring from his position at Aarhus University, he continued to pursue his research interests and expand his international publishing activities as an independent researcher. He characterized himself as an "outsider" to both established engineering and traditional disciplinary academic settings, taking pride in this marginal position. He argued that "[o]ne of the many benefits of an 'outsider' perspective is that outsiders may be able to bring in new perspectives, maintain a greater degree of detachment, and be in a better position to disclose taken-for-granted assumptions…" (Christensen, 2012, 1). His many achievements proved his point right.

Steen will be remembered by many colleagues for his enthusiasm, curiosity, tenacity, and generosity. He possessed a unique blend of appreciation and respect for knowledge and achievement, coupled with a levelheaded and unimpressed approach to the academic establishment. This rare combination enabled him to stay curious and reach out to leading researchers as well as neophytes in the research field. What mattered to Steen were ideas and people, not institutional conventions or the promotion of his own position in academia. He cherished exploring authorships and intellectual traditions as an inspiration for better understanding what set them apart from other traditions, and he was thrilled if insights from one tradition could enrich another. Exploring, challenging, and questioning the binary opposites of engineering and the humanities, the technical and the social, the practical and the theoretical, was a stimulating intellectual endeavor for Steen. The loss of Steen will be mourned by many who have an interest in better understanding the two cultures of engineering and the humanities, but his contributions will support new generations of scholars in continuing important discussions in the field of the philosophy of engineering and technology through the many books he brought to life and the intellectual community he established.

Anders Buch

References

Buch, A. & Christensen, S. H. (Eds.), *Bildung for engineering education and practice: A new agenda*. Springer.

Christensen, S. H. (2012). *Engineering in context: Structural, cultural, social, and epistemological tensions in engineering education*. PhD thesis, Faculty of Engineering and Science, Aalborg University.

Christensen, S. H., Delahousse, B., & Meganck, M. (Eds.) (2007). *Philosophy in engineering*. Academica.

Christensen, S. H., Delahousse, B., & Meganck, M. (Eds.) (2009). *Engineering in context*. Academica.

Christensen, S. H., Mitcham, C., Bocong, L., & An, Y. (Eds.) (2012). *Engineering, development and philosophy: American, Chinese, and European perspectives*. Springer

Christensen, S. H., Didier, C., Jamison, A., Meganck, M., Mitcham, C., & Newberry, B. (Eds.) (2015a). *International perspectives on engineering education; Engineering education and practice in context, Volume I*. Springer.

Christensen, S. H., Didier, C., Jamison, A., Meganck, M., Mitcham, C., & Newberry, B. (Eds.) (2015b). *Engineering identities, epistemologies and values*: *Engineering education and practice in context, Volume II*. Springer.

Christensen, S. H., Delahousse, B., Didier, C., Meganck, M., & Murphy, M. (Eds.) (2019). *The engineering-business nexus: Symbiosis, tension and co-evolution*. Springer.

Christensen, S. H., Buch, A., Conlon, E., Didier, C., Mitcham, C., & Murphy, M. (Eds.) (2022). *Engineering, social sciences, and the humanities: Have their conversations come of age?* Springer.

Jamison, A., Christensen, S. H., & Botin, L. (2011). *A hybrid imagination. Science and technology in cultural perspective*. Morgan & Claypool.

Contents

Chapter 1
Introduction: Contesting Narrow Conceptions of Engineering Education and Practice Through the Concept of *Bildung*

Steen Hyldgaard Christensen and Anders Buch

Abstract This chapter introduces the theme and the chapters of the anthology. Inspired by the role the concept of *Bildung* has come to play in science education the chapter first seeks to provide responses to two fundamental questions, namely (1) what is the purpose of engaging in research on *Bildung* generally, and (2) what is the purpose of introducing the concept of *Bildung* in engineering education and practice particularly. For that purpose authors from anglo-saxon, skandinavian, francophone, and Chinese cultures discuss the relevance and challenges of this venture. The second part of this chapter briefly presents each contribution in its own right.

Keywords Introduction · *Bildung* · Crisis · Education · Engineeering · Challenges · Ethics

1.1 Why *Bildung* and Why Now?

At first sight, the concept of *Bildung* appears odd, irrelevant, untimely, overly complex, archaic, and a floating signifier for diverse and potentially contradicting visions for education, self-cultivation, and personal development. Furthermore, the concept of *Bildung* is naturally associated with an elitist German culture closely connected to German idealism and the romantic movement originating at the beginning of the nineteenth century in Jena. This particular culture is conjoined with Wilhelm von Humboldt and his reform of the Prussian educational system in 1809.

S. H. Christensen
Independent researcher, Herning, Denmark

A. Buch (✉)
VIA University College, Valby, Denmark
e-mail: buch@via.dk

A. Buch, S. H. Christensen (eds.), *Bildung for Engineering Education and Practice*, Philosophy of Engineering and Technology 49,
https://doi.org/10.1007/978-3-031-86581-7_1

For proponents, its allure lies in the fact that it refers to an institution dedicated to knowledge and the initiation of students to cultural traditions and values through research, scholarship, and art. It stresses the unity of research and education free from any pressure from the state or the market. Traditional visions of *Bildung* see its mission as distinct from vocational or professional training and education. Its objective is to offer each student the greatest possible freedom and diversity of experience to allow them to construct their own learning, without any utilitarian, vocational, political or other ulterior motives. This ideal is reflected in Emanuel Kant's treatise The Conflict of the Faculties. Here, Kant argued that as opposed to the utility-oriented higher faculties of theology, law, and medicine, the philosophical faculty must be free because "it must answer for the truth of the teachings it has to adopt or even allow, must be conceived as free and subject to laws only given by reason, not government" (Kant, 1798, cited in Geer Hammershøj, 2019, p. 162). Since Humboldt and Kant promoted their visions of higher education, fundamental changes may be observed in many higher education systems in Europe and beyond. Educational pursuits are no longer perceived as standalone activities; instead, they are often regarded as a means for economic growth and the prosperity of the business community. These tendencies have been labeled academic capitalism and academic neoliberalism. The changes have been aptly described by Klarissa Lueg and colleagues:

> Universities face increasing national and international competitiveness for status and financial resources and respond to this with different policy measures and strategic organizational development. These changes go along with the idea that the production of internationally visible top-level research and teaching is best stimulated by market-based competition and is to be professionally managed. (Lueg et al., 2020, p. 267).

The crucial question, therefore, is why and with what effects attempts to reinvigorate the concept of *Bildung* have taken place in contemporary discussions about the role and purpose of higher education in society. The concept of *Bildung* often emerges prominently as a response to a crisis when the existing educational system is no longer capable of fulfilling its promises to society (cf. Heikkinen et al., 2023). *Bildung* thus serves as a counter or problematizing concept in relation to the current configuration of the educational system, or simply as a marker of crisis, as observed in discussions on higher education in many countries. The recent centrality of *Bildung* in the ongoing education debate should therefore be understood as a critical response vis-à-vis the increasing instrumentalization of educational systems overall. With the neoliberal conception of national states' competitiveness, higher education no longer seems to prioritize *Bildung* in terms of nurturing individual students. Rather, the purpose of higher education is viewed as supplying skills and knowledge to professionals, aligned with the notion of employability drift, vocational drift, and practical drift (Tight, 2015). According to critics, however, higher education is currently left in a state of crisis due to the influence of neoliberal ideology.

1.2 Why *Bildung* for Engineers?

Joseph Herkert characterizes the prevailing engineering culture and its associated professionalism in an ideal typical fashion as follows:

> The prevailing engineering culture is readily recognized from both inside and out. Engineers are no-nonsense problem solvers, guided by scientific rationality and an eye for invention. Efficiency and practicality are the buzzwords. Emotional bias and ungrounded action are anathemas. Give them a problem to, specify the boundary conditions, and let them go at it free of external influence (and responsibility), if problems should arise beyond the work bench or factory floor, these are better left to management or politicians (Herkert, 2001, p. 410).

Important to note here is that engineering problem-solving is traditionally viewed as primarily reliant on the engineering sciences and/or the natural sciences. The conventional argument suggests that because the engineering curriculum must encompass a broad range of technical subjects and disciplines, little room is left for broader social and ethical considerations. However, integrating technical and scientific subjects with social issues seems unavoidable for practicing engineering in the real world. James Trevelyan notes that:

> Studies have shown, for example, that technical problem-solving in practice largely relies on social interactions with other people. For example, a study of problem-solving in the automotive industry demonstrated that gaining alignment among different stakeholders on the nature of the problem was the main factor that distinguished success from failure, in essence a multi-party negotiation performance. (Trevelyan, 2025, Chap. 11 of this volume).

A common approach to bridging the gap between technical and broader societal aspects has been the inclusion of standalone subjects on ethics and social responsibility into an already packed engineering curriculum. However, standalone engineering ethics courses, often taught by faculty with backgrounds in the humanities or social sciences, exemplify problematic attempts to achieve integration. Instead, the integration of ethics and questions of social justice into the engineering curriculum must be rooted in technical problem-solving as it is practiced in the field. This approach speaks to the need for engineering education to become more relevant and responsive to the challenges engineers face when solving problems.

In considering their designs and problem-solving approaches, engineers must incorporate procedural knowledge, including assessing benefits and costs in terms of who bears the burdens and who receives the benefits, whether financial or otherwise (Cumming-Potvin & Currie, 2013). An illustrative example provided by Gravel and Svihla (2020) is the construction of a bridge connecting two communities. This technical undertaking requires designers to engage with the respective communities to achieve success. Addressing concerns such as noise, traffic flow, aesthetics, and the impact on the communities is essential for providing effective solutions.

Given that engineering is a heterogeneous sociotechnical enterprise, the inadequacy of traditional engineering perspectives, as outlined in Herkert's ideal type of engineering professionalism, becomes apparent. Something needs to bridge the gap between technical and broader societal aspects and stand-alone social science and ethics courses apparently does not do the job. Might (re)considering the concept of *Bildung* be helpful in this regard? Can the 'two cultures' (Cheville, 2022) of engineering and the social sciences and humanities (SSH) be conjoined in helpful ways by (re)considering the ideas of *Bildung* in new ways? These are the central questions that are discussed in this anthology.

1.3 The Historical Roots of *Bildung*

The historical roots of the concept of *Bildung* can be traced as far back as the Greek notion of *Paideia.* Sven Erik Nordenbo concisely captures the ancient Greeks notion of *Bildung* as Paideia:

> From the Greek point of view, *Bildung* is precisely *neither* private (that is, about personal or commercial usefulness), *nor* about public utility. The cultivation of man on his own definition (man's *Bildung*) manifests itself instead as a general structure, within which the individual sphere and the public or general sphere are in harmony. Thus, thinking about *Bildung* undermines all attempts to limit man to specific social functions alone. (Nordenbo, 2003, p. 30)

Paideia centered on the notion of *Arete*, encompassing excellence, virtue, and fulfillment, with Arete herself representing a goddess of virtue, goodness, and knowledge. Although not a direct source of inspiration for the development of traditional notions of *Bildung*, similar notions of virtue and self-improvement can also be found in China, based on Confucianism (cf. Zhu & Sun, 2025, Chap. 12 in this volume).

Historically, the concept of *Bildung* can be conceived as an amalgamation of Greek and Christian ideas of virtue and piety. The concept of *Bildung* originates from the German term 'Bild' (image), suggested by the Christian Dominican theologian and philosopher Johannes Eckardt, or just Meister Eckhart (1260–1328), to convey the notion that thinking of oneself should be "as made in the image of Christ." (Horlacher, 2016, p. 8). But it was until the eighteenth Century that the notion of *Bildung* found its classical expression. It was Johan Joachim Winckelmann, a prominent German art historian, who, in 1755, articulated his art program in his book *Gedanken über die Nachahmung der Griechischen Werke in der Malerei und Bildhauerkunst*, inspiring several German authors, including Wilhelm von Humboldt. Winckelmann believed that the creation of a distinctive German culture could only occur in opposition to the Latin-Roman culture represented by France. For Winckelmann, 'Edle Einfalt und stille Größe' (noble simplicity and quiet grandeur) embodied the spirit of Greek art and German grandiosity. In light of this, as noted by Rebekka Horlacher, specific ideas of education and associated values emerge from concrete historical contexts. In comparison to the historical contexts from which the German concept of *Bildung* and the French concept of *civilization* emerged, she observes:

> The emphasis on *Bildung* follows from the fact that many German intellectuals thought of themselves as "backward" by comparison with France, a country that regarded itself a great "nation of civilization". The Germans leveraged the concept of *Bildung* as a means of distancing themselves from the French idea of civilization – an idea they criticized as centralist, courtly, alienated, and morally suspect – claiming "culture" and *Bildung* as uniquely German and, in truth, superior. (Horlacher, 2016, p. 45).

However, the classical German concept of *Bildung* was first fully articulated in the writings of romantic philosopher Johann Gottfried Herder in his seminal *Journal of My Voyage in the Year 1769*. Central to Herder's idea of *Bildung* is the belief that human development and education are deeply intertwined with cultural and historical context. Unlike Enlightenment thinkers who emphasized universal principles, Herder emphasized the uniqueness of 'lived' cultures. He rejected the notion of a single, fixed standard of civilization, instead advocating for an appreciation of the concrete historical human experiences of what individuals consider real and meaningful. Moreover, Herder believed that true *Bildung* requires a deep understanding of one's cultural heritage and a sense of historical consciousness, and he emphasized the role of education in fostering this connection to one's roots and nurturing a sense of belonging to a larger cultural community. By embracing the richness and complexity of human culture, he envisioned *Bildung* as a means of fostering empathy, tolerance, and appreciation for the diversity of human experience. The cultivation of moral and aesthetic sensibilities in human development was thus considered a basic prerequisite for the formation of both self and society.

The German concept of *Bildung* was further developed in Georg Wilhelm Friedrich Hegel's philosophy. He considered *Bildung* as a process of self-realization and self-development, where individuals strive to actualize their potential and achieve spiritual fulfillment. He considered this process to involve the cultivation of reason, morality, and aesthetic sensibilities through engagement with cultural and intellectual pursuits. *Bildung*, for Hegel, is thus not only a matter of acquiring knowledge or skills but rather a holistic transformation of consciousness and character. Neither should it be considered as a process only relevant for the development of individuals but also a collective phenomenon, a shaping of culture, institutions, civilization and of the cosmos itself over time (Herdt, 2019).

Humboldt, like Herder and Hegel, acknowledged the important role that formal education played for the *Bildung* of individuals and society. He believed that education should not only transmit knowledge but also foster critical thinking, moral autonomy, and a sense of civic responsibility. Even though Humboldt's ideas of *Bildung* was primarily focused on higher education and research (studium generale)—the sole privilege of the few and wealthy citizens of the Prussian bourgeoisie—he also considered *Bildung* as relevant for secondary education and to some extent for vocational education. He argued that vocational training should not be isolated from the liberal arts and sciences but should be complemented by a well-rounded education that fostered critical thinking, creativity, and a sense of civic responsibility—much in line with the central thoughts of the European Enlightenment philosophy.

1.4 How Ideas of *Bildung* Have Travelled

Even though ideas of *Bildung* are largely being ignored in present day mainstream higher education—and particularly in engineering education—they have survived, mutated, and merged with other intellectual strands of thought as counter-narratives throughout the two centuries that have passed since their birth in Jena. Likewise, the ideas have travelled to different academic disciplines and outside the Germanic cultural sphere to different regions. Today, new and modified visions of *Bildung* are being voiced in different contexts as an antidote to the shortcomings of both elementary and higher education and the cultural and political decline brought about by consumerism, globalism, and populism.

Jesper Sjöström and his colleagues (e.g., Sjöström et al., 2017; Sjöström & Eilks, 2018) have explored how the concept of *Bildung* is used in the international science education literature and they have identified different conceptualizations of the *Bildung* idea that are presently promoted in various contexts. Building their classification on previous research, they identify at least five different versions of *Bildung* in contemporary international science education literature: a classical *Bildung* tradition, a liberal education tradition, a Scandinavian *folk-Bildung* tradition, a democratic education tradition, and a critical-hermeneutic *Bildung* tradition.

The classical *Bildung* tradition takes inspiration from Humboldt's original vision, that the primary ambition of education should be to stimulate the holistic development of individuals. It encompasses intellectual, moral, and cultural dimensions of education, aiming to cultivate individuals' capacities for critical and systematic thinking, ethical awareness, and appreciation of a cultural canon. The Humboldtian conception of *Bildung* thus emphasizes the importance of academic freedom, autonomy, and the integration of teaching and research within the university system. This classical notion of *Bildung*, however, is often critiqued for being elitist, intellectualist and biased towards the nineteenth century nationalist German cultural context, to reproduce existing social hierarchies, to disregard the practical realities, and praising repressive conservative values. Critics have thus aimed to modify the concept in various ways.

One transformation can be found in the Anglo-American tradition of liberal education. This tradition emphasizes the cultivation of intellectual curiosity, breadth of knowledge, and critical thinking skills. Following the fundamental *sapere aude* parole of the Enlightenment ideal, liberal education aims to provide students with a broad-based education that prepares them for active citizenship, lifelong learning, and meaningful participation in society. It values interdisciplinary exploration, open-minded inquiry, and the pursuit of truth and understanding across diverse fields of study. Although the liberal arts tradition can be regarded as a reaction to the classical notion of *Bildung*, it shares its elitist and intellectualist characteristics. In a postmodern setting where faith in universal truths and knowledge have been shattered, the ideals of liberal education have, however, been reconsidered to stress the de- and re-constructive cultural narratives to edify character traits. For Richard

Rorty (1998), edification involves the process of self-improvement or personal growth through engagement with cultural, intellectual, and moral resources.

In the Scandinavian context the notion of *Bildung* has been developed in the direction of *folk-Bildung*. This version of *Bildung*, rooted in Scandinavian educational traditions, emphasizes the importance of community, equality, and social cohesion in education and public life. *Folk-Bildung* focuses on the development of individuals within the context of their communities, fostering solidarity, mutual respect, and a sense of belonging. It values collaborative learning, experiential education, and the integration of practical skills with theoretical knowledge. Folk-*Bildung* aims to empower individuals to actively participate in society and contribute to the common good. The Scandinavian *Folk-Bildung* tradition seeks to balance individualist ideas of the Enlightenment-project with the communitarian ideas of romanticism but avoiding the elitism of both strains of thought.

Another, but slightly reminiscent to Scandinavian *folk-Bildung*, avenue has been paved by educators and philosophers inspired by John Dewey's thinking about the role of education and democracy. This version of *Bildung* emphasizes the principles of democracy, equity, and social justice in education. It views *Bildung* as a means of promoting democratic values, citizenship education, and transformation of social institutions and ways of living. Democratic education values inclusive decision-making, student empowerment, and the recognition of diverse voices and perspectives. It aims to create learning environments that promote equality, respect for human rights, and active participation in democratic processes.

The final version of *Bildung* observed by Sjöström and his colleagues sets out to transform the classical conception of *Bildung* to encompass ideas from critical and hermeneutic scholarship. Critical-hermeneutic *Bildung* draws on critical (Marxist) theory and hermeneutics to explore the role of education in fostering critical consciousness, self-reflection, and social critique. It encourages students to question and challenge prevailing norms and power structures, engage in dialogue with diverse perspectives, and critically analyze social, cultural, and scientific discourses. It aims to empower individuals to become active agents of change, challenging oppression, inequality, and injustice in society. Taking inspiration from the Frankfurt-school, Hans-Georg Gadamer's philosophy and Paul Ricæur it has become an inspiration for educators mainly in contemporary Germany and parts of Scandinavia.

The mapping of Sjöström and his colleagues of the different roads that have been taken by educators and philosophers in transforming the concept of *Bildung* does not pretend to exhaust and give credit to nuances in positions in the contemporary international science education literature. Likewise, many crossovers between the five versions can be observed. However, the framework provided by Sjöström and his colleagues gives us a useful roadmap for exploring the landscape of contemporary discussions about *Bildung*. For our purposes, it has given inspiration to further explore if and how the concept of *Bildung* is useful for rethinking engineering education and professionalism in new ways.

1.5 Ideas of *Bildung* in Contemporary Engineering

The rationale for this anthology grew out of our previous research on engineering education and practice. Over the last two decades we have—in collaboration with a wide-ranging international community of scholars—explored various themes and aspects relating to the philosophy of engineering and technology (cf. Christensen et al., 2007, 2009, 2012, 2015a, 2015b, 2019, 2022). In this volume we wish to continue the exploration by reflecting on the prospects of reforming engineering education and practice. Our questioning has lashed on to notions of *Bildung* to inquire how this and related ontological, epistemological, and ethical conceptions of the telos and raison d'être of engineering is articulated, evaluated, and enacted by contemporary engineering scholars. Inspired by the work of Sjöström and his colleagues, we challenged a diverse group of international engineering scholars to critically reflect on the relevance and appropriateness of introducing conceptions of *Bildung* into engineering education and practice: In a time of crisis, do engineers need *Bildung*? Should engineering education be broadened and how? How should we think about engineering and engineering education in and across cultural contexts? These and related questions have guided the discussions in the 12 chapters that follows.

The authors of the chapters are all engaged in teaching engineering and/or researching engineering education and practice, but their experiences are rooted in different cultural contexts and practices. The perspectives on engineering education and practice are proposed by authors from different cultural spheres, including the Anglo-Saxon, Scandinavian, Francophone, and Chinese cultural spheres. To some extent the perspectives are congruent with and overlaps the different approaches listed in Sjöström et al.'s mapping, but new and specific dimensions, issues, and ideas are also introduced in considering the relevance of *Bildung* for engineering education and practice—e.g., in the contributions of the French and Chinese authors. All authors agree that engineering education and engineering practices are in need of reforms to face the enormous challenges that humanity face. They also agree that reconsidering issues of *Bildung* is a timely and appropriate endeavor. It seems mandatory to broaden the engineering mindset beyond narrow and delimited technical concerns to meet the challenges. But the authors' visions suggest different inroads for identifying and dealing with the challenges. The authors thus de- and re-construct the notion of *Bildung* in various ways to find appropriate approaches that seem promising in relation to the specific cultural contexts that they have encountered in their teaching and research. The remainder of this introduction will briefly present each contribution to the anthology.

1.6 Presentation of the Chapters

Part I of the anthology comprises five chapters where the authors discuss the need for *Bildung* in engineering.

Chapter 2, *Rethinking Engineering Professionalism through the Concept of Bildung*, authored by Anders Buch and Steen Hyldgaard Christensen (Buch & Christensen, 2025a), sets out to rethink the defining telos of engineering professionalism. The authors note how recent developments in science and environmental studies coalesce with the needs for reform in engineering education. They argue that such reform should be based on an understanding of engineering as a heterogeneous sociotechnical enterprise in which challenges and solutions are distributed across material, economic, cultural, political, and technical domains. As the authors wish to take into account the challenges facing humankind in the Anthropocene era, they argue that the telos of engineering, traditionally defined in the US as "the art of directing the great sources of nature for the use and convenience of man," should be changed. They take inspiration from John Dewey's understanding of democratic education to do so. They propose that engineering should be reassessed as a problem-solving endevour that seeks to ameliorate living conditions. According to the authors, a Deweyan framework enables acknowledging that grand challenges such as global warming, pandemics, decreasing biodiversity, economic and social polarization, racism, information and cyberwars, and cybercrimes are complex problems that cannot be solved through technical fixes. To find viable solutions for these grand problems engineers need to reconsider how these problems are framed and broaden their conceptions of the nature of the problems. Yet, barriers and obstacles for such ventures come along two axes, a sociocultural and an institutional one, they argue. The first relates to how engineering identities are reproduced. The second axis concerns how ecologies of practices including teaching, learning, assessment, curriculum design, and classroom management reproduce traditional institutional frameworks.

Chapter 3, *Bildung under Conditions of Engineered Destruction*, by Mitcham (2025), is partly a response to the concept of *Bildung* proposed by Buch and Christensen for rethinking engineering professionalism, but the chapter has wider aims. Mitcham addresses what he calls apocalyptic engineering *Bildung*. The chapter starts by asking what it might mean to aspire to the realization of a specifically engineering *Bildung* ideal under conditions in which engineering itself has become a major source for social-political and planetary disorder. The first part of Mitcham's response challenges assumptions in traditional conceptions of *Bildung*, namely that the world is well ordered, and the assumption that engineering is in fact able to make things better. Mitcham contests both of these views. However, Mitcham also approvingly notes that Buch and Christensen's attempt to reconstruct conceptions of *Bildung* and engineering do not stipulate these assumptions. Lastly, Mitcham propose that under conditions of potentially apocalyptic engineering, revisiting the pre-Enlightenment meaning of *Bildung* where humans were seen as sinful creatures would not be misplaced. He urges us to learn from the German struggle to overcome the autrocities of the Nazi-regime by *Vergangenheitsbewältigung* and suggests that engineers need to acknowledge and repend the misdeeds of the profession in order to proceed in responsible ways.

In Chap. 4, *Bildung, Destruction, and Reconstruction: A Rejoinder to Carl Mitcham*, Buch and Christensen (Buch & Christensen, 2025b) reflect on Mitcham's proposals. They acknowledge the relevance and timeliness of Mitcham's cautionary observations, but hesitate to follow the pessimistic and retrospective approach of his suggestions. Instead, Buch and Christensen maintain that *Bildung* must be future-oriented and address the precarious nature of our existence.

In Chap. 5, *From the Individual to the Collective Dimension of Bildung in Engineering Education through Challenge-Based Learning*, Diana Adela Martin, Andreas Spahn, and Gunter Bombaerts (Martin et al., 2025) address the questions: How can we understand *Bildung* as a goal in contemporary engineering education, and how does Challenge-Based Learning (CBL) serve as a pedagogical approach to accomplish it? The authors trace the historical interpretations of the concept of *Bildung*, and they observe that the ideal of *Bildung* presently is reflected in the discourse about the need to develop holistic and well-rounded graduates capable of exercising their social responsibility and applying their technical skills to societal issues. In the classroom, this translates into a focus on ethics, virtues, empathy, social justice, decolonization, and global and cultural awareness. However, the authors also note that in the institutionalization of *Bildung*, both individual agents including students, teachers, and administrators, as well as collective agents such as universities, schools, firms, and community groups, should be able to learn based on respect for their self-realization, motivation, understanding, and autonomy. The authors argue that the pedagogy of CBL brings real-life problems to the classroom based on the activities of real-life stakeholders such as external research and development centers, startups, companies, NGOs, citizen associations, local communities, etc. The chapter offers inspiration to educatiors who seek to foster *Bildung* in ther teaching by offering practical examples.

In Chap. 6, *Thinking about Engineering: Between Bildung and Citizenship,* François Prouteau, Nathanaël Wallenhorst, Renaud Hétier (Prouteau et al., 2025) propose to integrate the concept of *Bildung* with the concept of *Citizenship.* Articulating this integration, they constrain themselves to a paradigmatic, theoretical, and conceptual level. They depart from the contemporary ecological awakening movement *Pour un réveil écologique,* spearheaded by graduates from French elite engineering schools. They argue that a helpful resource to conceptualize this emerging movement is the scholarship of the Franco-German ecomist Christian Arnsperger, who introduces the term "existential citizens". This term denotes citizens who make radical life choices and voluntarily break with their capitalist existence. The authors conceive of their chapter, belonging to the field of political anthropology, as a parallel endeavor to that of Sjöström et al. (2017). Accordingly, they argue that citizenship can accommodate both technical ability employed to benefit a City wishing for development, progress, and continual improvement as well as the human concern that *Bildung* embodies. In elaborating the meaning of exercising citizenship, they consider the political theory developed by Hannah Arendt. In doing so, the authors use Arendt's concepts of *Work, Labor, and Action.* They stress that under *Work,* Arendt uncovers the risk implicit in confusing "education" with mere training in a narrow set of technical skills. *Labor* is about biological needs but has also, in ancient times, been associated with people deprived of the right to take part in political decisions. The opposite of this is *Action,* which refers to the fulfilment of life in the political sphere.

Part II of the anthology comprises four chapters in which the authors discuss intersections between *Bildung* and efforts to broaden engineering education.

Eddie Conlon, in Chap. 7, *Bildung: Conformity or Transformation?,* (Conlon, 2025) from a Marxist perspective, sets out to explore how the concept of *Bildung* can be used to build an argument for a transformational approach to engineering education. This is deemed necessary because engineering education has become captive to several agendas imposed by corporate actors. As a result, the fundamentally exploitative nature of capitalism is rarely addressed in dominant educational discourses. This is leading students to internalize the demands of the methods of production as self-evident. Conlon goes further and argues, that while neo-liberalism represents the most aggressive form of capitalism, the captivity of education to the economic system is not a new problem. Calling attention to Paulo Freire's *Pedagogy of the Oppressed,* Conlon maintains that education should become the practice of freedom as there is no such thing as a neutral education. If this is not happening, education will be used to facilitate the integration of the younger generation into the logic of the prevailing system and bring about conformity to it. Hence, transformative learning requires a qualitative shift in our assumptions, habits of thoughts, a changing understanding of power relations, and a sense of possibilities for social justice and change. Moreover, a system based on the exploitation of labor, constant growth in pursuit of profit, and universal commodification will have deleterious consequences for the planet and the people living on it. Finally, Conlon critically comments on the significant role employers play in Irish engineering education as participants in the accreditation system, and in some institutions even as external examiners.

Karl Palmås, in Chap. 8, *Bildung, and how that Concept Sits with Traditional Notions of Teaching Students "How to Think" like Engineers,* (Palmås, 2025) investigates the meaning of *Bildung* in the context of engineering. He comprehends the *modus operandum* in engineering practice as breaking large and complicated problems into manageable components, solving each lower-level problem one by one, thus deriving a higher-level resolution. In considering how to think, the chapter hints to what American novelist David Foster Wallace calls the "standard liberal arts analysis" which breeds tolerance and an appreciation of of diversity of beliefs. The chapter proceeds by examining central themes in the literature on *Bildung* in engineering education focusing on the five different understandings of *Bildung* outlined by Sjöström et al. (2017). Additionally, it introduces two important figures whose work was central to how the concept of *Bildung* was introduced and re-interpreted in Sweden during the nineteenth century, namely Erik Gustav Geijer and Carl Jonas Løve Almquist. Their work highlight the way to come to terms with the classics. Palmås emphasizes, that for engineering students, creative readings are more productive than the close readings conducted within the humanities because it allows students to use their education to figure out their own political or ethical positions. Moreover, in presenting engineering education as an inheritor of the ideals of classical *Bildung,* it is possible to clarify the rules of engagement between teachers and students and for students to transcent teachers' vested oppinions and interpretations.

Sarah Appelhans and Alan Cheville, in Chap. 9, *Appropriateness and Value: Cultivating Bildung Mindsets Amongst Undergraduate Engineers in a Third-Year Design Course,* (Appelhans & Cheville, 2025) present findings from a course for undergraduate engineering students aimed at evaluating the societal impacts of engineering problem-solving. The course took place in an electrical and computer engineering department located at an American liberal arts university. Their design course was aligned with an understanding of engineering as a heterogeneous socio-technical enterprise. The primary aim was to challenge dominant epistemologies that frame engineering disciplines as merely applied science. In addressing this framing, the authors also discuss the intellectual captivities imposed by social forces that seek to use technology for their own ends. They further note that the action-oriented nature of engineering work, involving application, problem-solving, and the creation of technology, positions engineering at the "sharp, pointy end of *Bildung"*. In their view, *Bildung* offers a better approach to support individual agency and foster moral and social responsibility as core components of engineering education. Regarding their research design, the authors describe the methods, participants, and data collection as follows: An ethnographer was embedded in the design class to understand the generative tension between liberal arts and engineering as it intersects with the key tenets of *Bildung,* as noted in student reactions. Data were collected over two semesters of a third-year design course in the 2021–2022 academic year. The authors conclude that by encouraging student to reflect on their values and consider their impact on society, *Bildung* has the potential to challenge the cultures of technical narrow-mindeness and cultivate a a more socially-conscious, morally responsible discipline.

In Chap. 10, *Usefulness and Bildung: Engaging Students in a Bildung-oriented Engineering education,* Ronny Kjelsberg and Magnus Strøm Kahrs (Kjelsberg & Kahrs, 2025) begin with the observation that for future engineers, the concept of "usefulness" expands to include normative connotations beyond mere applicability. They investigate how societal perspectives on usefulness align with engineering students' reported motives for pursuing an engineering education. By comparing the framing of a bachelor program for construction engineering in 2019 and 2023, they note the inclusion of broader societal perspectives in the latter, implying the relevance of a *Bildung* perspective.. The authors generate empirical data from responses to the question: Why do you want to study engineering? The survey covers the years 2019–2022 and includes 16 themes. Regarding the main theme of usefulness, the authors observe a strong motivation among engineering students towards practical, concrete, and useful contributions to society, including addressing issues like climate change and technological development in general. They conclude that for a *Bildung*-oriented engineering education to succeed, it must encompass various forms of usefulness and relevance centered around engineering students' perspectives.

Part III of the anthology comprises three chapters in which the authors discuss *Bildung* across Cultural Contexts.

In the opening Chap. 11, *Bildung and Education for Engineering Practice,* (Trevelyan 2025) James P. Trevelyan notes that all teaching in medical schools, beyond the basic sciences, an introduction to pharmacology, and social and population sciences, relies on instructors who practice medicine on a daily basis. In contrast, few teachers in engineering faculty have present-day experience from engineering practice.. Trevelyan argues that engineers practice their technical expertise within complex social and organizational contexts. However, this is not reflected in engineering education. He argues that professional engineering education could be significantly enriched by helping students to become competent social actors through understanding their own personal development by integrating ideas from the German concept of *Bildung.* Integrating the concept of *Bildung* will prepare students to understand engineering workplace practices, help them to develop a stronger sense of identity and agency as engineers, and help them understand why costly engineering failures occur and how they might be prevented in the future. Agency is required, as technical problem-solving in practice relies on social interaction with others, as evidenced by a study of problem-solving in the automotive industry. The study indicated that achieving alignment among different stakeholders on the nature of the problem was a primary factor distinguishing success from failure.

Qin Zhu and Yixiang Shawn Sun in Chap. 12 *Toward a Self-Cultivation Approach to Engineering Education: A Confucian Perspective* (Zhu & Sun, 2025) starts from their percertion that a self-critical and self-reflective dimension have largely been lacking in dominant approaches to the formation of future engineers. Their awareness of a perennial problem in engineering education underscores philosopher Carl Mitcham's observation that the true grand challenge for engineering is self-knowledge in terms of the ability to think reflectively and critically about the kind of world we wish to design, construct, and inhabit through our technologies. Confusian Self-Cultivation theory in combination with the German concept of *Bildung* offers resources for addressing this challenge. Confusianism in the authors' understanding comprises three key elements: (1) Deep reflection to refine one's character. (2) The practice of virtues and refinement of moral character should be conducted in one's social interactions with others. (3) Lifelong pursuit of knowledge and continuous self-improvement should constitute the epistemological basis for self-cultivation. Furthermore, the authors establish relationships between the German concept of *Bildung* and Confusian Self-Cultivation theory. In this respect they note that *Bildung* places strong emphasis on the signicance of knowledge as a means to achieve individual freedom or a more holistic self. The autors additionally propose outcomes that should be considered in ABET accreditation procedures. Ultimately, they argue that future research based on the above ideas should discuss how the idea of self-cultivation can help design better learning activities and assessment strategies, and last but not least a flourishing life for engineers.

In the last chapter of Part III *Humanities and Engineering Education for Gaston Berger from the Perspective of Bildung* Marie-Pierre Escudié (Escudié, 2025) propose to reread Gaston Berger's philosophical work from the perspective of *Bildung*. She unfolds the influence of Gaston Berger's thinking along two axes: (1) The institutionalization of his thought in 1957 through the creation of the French engineering school Institut National des Science Appliquées (INSA) in Lyon, and (2) The presentation of his educational principles. The emphasis of the chapter is put on the latter. The appearance of the INSA was intended as a rupture from the elitism of the traditional highly selective engineering schools such as Ecole Polytechnique and Ecole de Ponts et Chaussées. This was reflected in the social backgrounds of incoming students. In 1962 20.45% of the INSA students came from working class and peasant families. Furthermore, Berger did not see technology and culture as antithetical as technology can nourish culture, and culture can provide a sense of direction for technology when it creates anxiety. Hence, Berger's combination of of a professional school for engineers and a liberal arts education is consistent with this vision.

Finally, Anders Buch's concluding chapter *Bildung for Engineering Futures*, Chap. 14, synthezises the discussions to explore how the concept of *Bildung* becomes relevant as a response to evolving challenge perceptions in engineering education and practice.

References

Appelhans, S., & Cheville, A. (2025). Appropriateness and value: Cultivating *Bildung* mindsets amongst undergraduate engineers in a third-year design course. In A. Buch & S. H. Christensen (Eds.), *Bildung in engineering education and practice. A new agenda*. Springer.

Buch, A., & Christensen, S. H. (2025a). Rethinking engineering professionalism through the concept of *Bildung*. In A. Buch & S. H. Christensen (Eds.), *Bildung in engineering education and practice. A new agenda*. Springer.

Buch, A., & Christensen, S. H. (2025b). *Bildung*, destruction, and reconstruction – A rejoinder to Carl Mitcham. In A. Buch & S. H. Christensen (Eds.), *Bildung in engineering education and practice. A new agenda*. Springer.

Cheville, A. (2022). The C.P. Snow controversy. In S. H. Christensen, A. Buch, E. Conlon, C. Didier, C. Mitcham, & M. Murphy (Eds.), *Engineering, social sciences, and the humanities. Have their conversations come of age?* (pp. 95–110).

Christensen, S. H., Buch, A., Conlon, E., Didier, C., Mitcham, C., & Murphy, M. (Eds.). (2022). *Engineering, social sciences, and the humanities: Have their conversations come of age?* Springer Nature.

Christensen, S. H., Delahousse, B., Didier, C., Meganck, M., & Murphy, M. (Eds.). (2019). *The engineering-business nexus: Symbiosis, tension and co-evolution*. Springer Nature.

Christensen, S. H., Delahousse, B., & Meganck, M. (Eds.). (2007). *Philosophy in engineering*. Academica.

Christensen, S. H., Delahousse, B., & Meganck, M. (Eds.). (2009). *Engineering in context*. Academica.

Christensen, S. H., Didier, C., Jamison, A., Meganck, M., Mitcham, C., & Newberry, B. (Eds.). (2015a). *International perspectives on engineering education; Engineering education and practice in context* (Vol. I). Springer.

Christensen, S. H., Didier, C., Jamison, A., Meganck, M., Mitcham, C., & Newberry, B. (Eds.). (2015b). *Engineering identities, epistemologies and values: Engineering education and practice in context* (Vol. II). Springer Science+Business Media B.V.

Christensen, S. H., Mitcham, C., Bocong, L., & Yanming, A. (Eds.). (2012). *Engineering, development and philosophy: American, Chinese, and European perspectives*. Springer Science+Business Media B.V.

Conlon, E. (2025). *Bildung*: Conformity or transformation? In A. Buch & S. H. Christensen (Eds.), *Bildung in engineering education and practice. A new agenda*. Springer.

Cumming-Potvin, W., & Currie, J. (2013). Towards new literacies and social justice for engineering education. *International Journal of Engineering, Social Justice, and Peace, 2*(1), 21–37.

Escudié, M.-P. (2025). Humanities and engineering education for Gaston Berger from the perspective of *Bildung*. In A. Buch & S. H. Christensen (Eds.), *Bildung in engineering education and practice. A new agenda*. Springer.

Gravel, B., & Svihla, V. (2020). Fostering heterogeneous engineering through whole-class design work. *Journal of the Learning Sciences, 30*, 279.

Hammershøj, L. G. (2019). The perfect storm scenario for the university: Diagnosing converging tendencies in higher education. *Futures, 111*, 159–167.

Heikkinen, A., Lassnigg, L., & Molzberger, G. (2023). What crisis are we in actually? The (Nordic) *Bildung* hype in adult education. In *Re-Konstruktionen-Krisenthematisierungen in der Erwachsenen-Bildung* (pp. 179–192). Verlag Barbara Budrich.

Herdt, J. A. (2019). *Forming humanity. Redeeming the German Bildung tradition*. Chicago University Press.

Herkert, J. R. (2001). Future directions in engineering ethics research: Microethics, macroethics, and the role of professional societies. *Science and Engineering Ethics, 7*, 3.

Horlacher, R. (2016). *The educated subject and the German concept of Bildung. A comparative cultural history*. Routledge.

Kjelsberg, R., & Kahrs, M. S. (2025). Usefulness and *Bildung*: Engaging students in a *Bildung*-oriented engineering education. InA. Buch & S. H. Christensen (Eds.), *Bildung in engineering education and practice. A new agenda*. Springer.

Lueg, K., Graf, A., & Powel, J. J. (2020). Hegemonic university tales: Discussing narrative positioning within the academic field between Humboldtian and managerial governance. In *Routledge handbook of counter narratives* (pp. 267–280). Routledge.

Martin, D. A., Spahn, A., & Bombaerts, G. (2025). From the individual to the collective dimension of *Bildung* in engineering education through challenge-based learning. In A. Buch & S. H. Christensen (Eds.), *Bildung in engineering education and practice. A new agenda*. Springer.

Mitcham, C. (2025). *Bildung* under conditions of engineered destruction. In A. Buch & S. H. Christensen (Eds.), *Bildung in engineering education and practice. A new agenda*. Springer.

Nordenbo, S. E. (2003). *Bildung* and the thinking of *Bildung*. In L. Løvlie, K. P. Mortensen, & S. E. Nordenbo (Eds.), *Educating humanity: Bildung in postmodernity*. Blackwell Publishing.

Palmås, K. (2025). *Bildung*, and how that concept sits with traditional notions of teaching students "how to think" like engineers. In A. Buch & S. H. Christensen (Eds.), *Bildung in engineering education and practice. A new agenda*. Springer.

Prouteau, F., Wallenhorst, N., & Hétier, R. (2025). Thinking about engineering: Between *Bildung* and citizenship. In A. Buch & S. H. Christensen (Eds.), *Bildung in engineering education and practice. A new agenda*. Springer.

Rorty, R. (1998). *Philosophy and social hope*. Penguin Books.

Sjöström, J., & Eilks, I. (2018). Reconsidering different visions of scientific literacy and science education based on the concept of *Bildung*. In Y. J. Dori, Z. R. Mevarech, & D. R. Baker (Eds.), *Cogntion, metacognition, and culture in STEM education: Learning, teaching and assessment* (Vol. 24). Springer International Publishing. https://doi.org/10.1007/978-3-319-66659-4

Sjöström, J., Frerichs, N., Zuin, V.-G., & Eilks, I. (2017). Use of the concept of *Bildung* in the international science education literature, its potential, and implications for teaching and learning. *Studies in Science Education, 53*(2), 165–192. https://doi.org/10.1080/03057267.2017.1384649

Tight, M. (2015). The development and application in higher education research: The case of academic drift. *Journal of Education Administration and History, 47*(1), 84–99.

Trevelyan, J. P. (2025). *Bildung* and education for engineering practice. In A. Buch & S. H. Christensen (Eds.), *Bildung in engineering education and practice. A new agenda*. Springer.

Zhu, Q., & Sun, Y. S. (2025). Toward a self-cultivation approach to engineering education: A confucian perspective. In A. Buch & S. H. Christensen (Eds.), *Bildung in engineering education and practice. A new agenda*. Springer.

Christensen, Steen Hyldgaard († 23rd June 2024). Independent researcher. He held an MA in Danish Language and Literature and the History of Ideas from Aarhus University, and PhD in research on engineering education from Aalborg University. Until 2014 senior lecturer at Aarhus University, Herning. From 2014 until May 2023, he was an adjunct associate professor at Aalborg University. He was initiator, coordinator, lead editor, and co-author of several edited volumes on engineering education, and epistemology. Five of them are published in the Springer Nature (POET) series. He has published articles on engineering epistemology, culture, and higher education. His research interests included the dynamics in higher education restructuring, and research initiatives to broaden engineering education.

Buch, Anders. Docent at VIA University College, Denmark and affiliated researcher at Jönköping University, Sweden. He holds an MA in philosophy from University of Copenhagen and a BEd and PhD in education from Roskilde University. His empirical research area is focused on professions, professionalism, and technological expert cultures, and his theoretical approach is primarily inspired by Science & Technology Studies, Practice Theory, and classical American Pragmatism. He has published articles and books on knowledge, learning, education, professionalism, and the professional development of engineers. Since 2016 he has servedas chief editor of Nordic Journal for Working Life Studies.

Part I
Engineers' and Technological Experts' Need for *Bildung*

Chapter 2
Rethinking Engineering Professionalism Through the Concept of *Bildung*

Anders Buch and Steen Hyldgaard Christensen

Abstract Research on engineering education and practice indicates that engineers' understanding of the "social" is underprioritized and discursively enacted as soft rather than hard technical disciplines. Despite the resounding accomplishments, due to democratic insufficiency engineering projects have often contributed to the cultural impoverishment of our lifeworld. Hence, a *Bildung* perspective is needed in engineering degree programs; a *Bildung* perspective taking its point of departure in the democratic, cultural, and human challenges of society. The discourse on *Bildung* has primarily taken place in relation to the pedagogical professions, whereas the engineering profession has often been overlooked. In this chapter, we make up for this by considering an understanding of *Bildung* that is suitable for engineering degree programs. The chapter brings forth the social and ethical responsibility embedded in engineering work in democratic societies and calls to mind John Dewey's democratic understanding of ethics, politics, and education. The chapter projects how this understanding might offer an opportunity to reconstruct *Bildung* and rethink engineering professionalism and the very telos of engineering to contribute to the democratic and responsible resolutions of societal and technological challenges.

Keywords *Bildung* · Engineering professionalism · Engineering education · Heterogeneity · Epistemology · Pragmatism

A. Buch (✉)
VIA University College, Valby, Denmark
e-mail: buch@via.dk

S. H. Christensen
Independent Researcher, Herning, Denmark

A. Buch, S. H. Christensen (eds.), *Bildung for Engineering Education and Practice*, Philosophy of Engineering and Technology 49,
https://doi.org/10.1007/978-3-031-86581-7_2

2.1 Introduction

The German concept of *Bildung* has been contested. For some scholars, it is seen as an anachronism in postmodern times, whereas for others, it holds great appeal (see Løvlie & Standish, 2003 and Siljander et al., 2012 for an overview). For Miller (2020), the concept seems, on the one hand, to promise a rich culturally situated form of rationality and, on the other hand, to serve as a powerful means of critique suited for our postmodern condition. However, Miller argues that *Bildung* is nevertheless destined to fail because it has little relevance to most pupils/students, not least by discarding vocational preparation and the cultivation of practical judgment, which are a vital concern for engineering professionalism. As Miller sees it, *Bildung* mainly appeals to intellectuals who seek fulfillment in the life of the mind. In contrast to Miller's skepticism regarding the relevance of *Bildung* for professionalism, Hans-Georg Gadamer is more positive:

> Practical *Bildung* is seen in one's profession wholly in all its aspects. But this includes overcoming the element in it that is alien to the particularity, which is oneself, and making it wholly one's own. Thus, to give oneself to the universality of the profession is at the same time "to know how to limit oneself"—i.e., to make one's profession wholly one's concern. Then it is no longer a limitation. (Gadamer, 1969/2013, p. 18)

The German ideal of *Bildung* is customarily associated with the name Wilhelm von Humboldt (Konrad, 2012). Even though the meaning of the ideal is complex, Humboldt nevertheless managed to briefly define it as the interplay and transaction between man and world:

> It is the ultimate task of our existence to achieve as much substance as possible for the concept of humanity in our person, both during the span of our life and beyond it, through the traces we leave by means of vital activity. This can be fulfilled only by linking the self to the world to achieve the most general, most animated, and most unrestrained interplay. This alone is the yardstick by which each branch of human knowledge can be judged. (Humboldt, 2000, pp. 58–59)

Thus, Gadamer established a fruitful connection between the Humboldtian concept of *Bildung* and professionalism, considering that the potential of the concept goes through reformulation and actualization in relation to changing living conditions. From the beginning, this ideal exposed a tension between self-cultivation through education, and its function as an instrument of social distinction (Humboldt 1793/2017).

As pointed out by Alexandre Alves (2019, p. 1), the present contemplation of this educational ideal should be seen as a marker of contrast to educational practices based on the imperatives of the market and neoliberal management of human capital (see also Rohstock, 2012). Today, the German concept of *Bildung* has become visible as a threshold concept in science and environmental studies, where critical thinking, societal values, ethics, and social responsibility have been introduced as new knowledge domains (Hansen et al., 2000; Sjöberg, 2005; Sjöström et al., 2017). Consequently, it makes sense to see the concept of *Bildung* as closely related to an understanding of social justice that includes fair and just relations between the individual and society as measured by the distribution of wealth, opportunities, and privileges within society—issues featured prominently in debates on globalization, capitalism, climate change, technology, and democracy.

In this chapter, we set out to argue that rethinking engineering professionalism through the concept of *Bildung* would be beneficial for addressing technological challenges and broader socially pertinent problems on a global scale, such as global warming, pandemics, decreasing biodiversity, increasing economic and social polarization, racism, cyberwars and cybercrimes. Dealing with such an array of challenges will have bearings on understanding the very telos of engineering (cf. Buch, 2022). Discussions of *Bildung* in engineering pertain to issues in higher education and the political philosophy of engineering. Ultimately, we will argue that the issue revolves as a struggle over knowledge, relevance, and disciplinary boundaries in relation to engineering epistemology.

In the first section of this chapter, we start by diagnosing the present configuration of higher education and the priority given to STEM fields (science, technology, engineering, and mathematics). We point to influential features that have increasingly structured higher education in the Global North, and we argue that the concept of *Bildung* usually rises to prominence when a given configuration of the educational system is in a state of crisis (Christensen, 2019; Hammershøj, 2017). Thus, it is useful to see *Bildung* as a counter and problematizing concept vis-à-vis present configurations of educational systems.

The second section starts with the observation that engineering is a heterogeneous sociotechnical enterprise in which challenges and solutions are distributed across social, material, economic, political, and technical domains (Gravel & Svihla, 2020). As "*Bildung* in Engineering Education" is a floating signifier, Jesper Sjöström and his coauthors' (Sjöström et al., 2017, p. 165) identification of the following five different understandings of *Bildung* in STEM education is relevant to consider for broadening engineering education: (1) Wilhelm von Humboldt's classical *Bildung*, (2) Anglo-American liberal education, (3) Scandinavian *folk-Bildung*, (4) democratic education, and (5) critical-hermeneutical *Bildung*. Considering the negotiated nature of engineering professionalism, together with the heterogeneous characteristics of its practices, we argue that option 4 has specific merits. Thus, the chapter points to this notion of *Bildung* and elaborates on John Dewey's understanding of democratic education to substantiate an ideal for the *Bildung* of contemporary and future engineering.

The third section unfolds elements in Dewey's philosophy that are helpful in this regard (Dewey, 2011). In doing so, it is necessary to address the very purpose of engineering: its telos. In the United States, the telos of engineering has traditionally been defined as "the art of directing the great sources of power in nature for the use and convenience of man" (Royal Charter of the Institution of Civil Engineers 1828). Considering the challenges facing humankind in the Anthropocene era, this telos must be reconsidered. We use Dewey's pragmatist lens to do so. Buch (2022) argued that Dewey's pragmatist attempts to reconstruct philosophy on the basis of a naturalist and humanist approach might serve as a productive vehicle to construct a new and more appropriate telos for engineering, that is, that engineering should be reconsidered as a problem-solving discipline that seeks to ameliorate living conditions. In the conclusion, we consider and discuss possible barriers and obstacles to reconstructing engineering along the lines of *Bildung* through democratic education.

2.2 Diagnosis: The Present Configuration of Higher Education and Future Sociotechnical Challenges

Since the 1980s, when the globalization of national economies took off, the neoliberal notion of the *competitive state* has served as the overall framework for educational reforms in nations in the Global North. Accordingly, higher education is seen as a vital infrastructure for innovation and competitiveness (Etzkowitz, 2008), where the primary functions of universities and higher education are to produce new knowledge for society and industry (Gibbons et al., 1994). Its corresponding political epistemology indicates the type of knowledge and competencies that should be targeted, combined with support for innovation and the establishment of an entrepreneurial culture in favor of STEM fields. The entrepreneurial university and neoliberal utilitarianism may be viewed as an assault on the Humboldtian university tradition of self-government and the very concept of *Bildung* (Rohstock, 2012). However, as the present configuration of educational systems within the former Humboldt tradition and beyond has not delivered on their promises (cf. Christopherson et al., 2014), the ensuing discourse on *Bildung* may be considered a marker of crises and a problematizing alternative concept vis-à-vis the present predicament of higher education.

Therefore, we argue that reclaiming and rethinking *Bildung* in new ways can be considered a viable strategy for critical educational scholars to counter the assault. In broader terms, we share the diagnosis that by using education policy as innovation and labor market policy, higher education has been thrown into a state of crisis. Increasingly, education is directed toward short-term employment needs in the present labor market, as defined by private sector requirements. As a result, more emphasis is placed on STEM fields, whereas the relevance of the social sciences and humanities (SSH) is set aside but becomes evident when considering the heterogeneous sociotechnical nature of what engineers do and accomplish.

By means of their professional knowledge, skills, and competencies, engineers contribute to creating not only the material and technological infrastructure of societies but also the professional, social, and cultural spaces we inhabit. The work of engineers contributes to shaping societies. Engineers advise, function as project managers, and contribute to the creation of new products, new knowledge, new work processes, and procedures of accountability. Furthermore, through technological innovation, they contribute to shaping the technological infrastructure for other professions such as those practicing within the health and education sectors.

In contrast to the descriptions of engineering as a simple problem-solving endeavor, emphasizing purely technical aspects, a more accurate understanding would describe engineering as a heterogeneous sociotechnical enterprise. Therefore, it is important to appreciate how the *social dimension* epistemologically flows into engineering problem-solving and what challenges and opportunities it provides. It is a well-established fact that engineering design and the application of technologies have political implications, shape social order, and are not value neutral (MacKenzie & Wajcman, 2005). Thus, Langdon Winner (1980) argued that artifacts have politics because they embody and promote certain values and ideologies that can have significant political implications. He suggests that the design and implementation of technologies can shape the social relations and power dynamics of societies in which they are used. According to Winner, technologies are not neutral; rather, they can reinforce or challenge existing power structures and embody particular social, economic, and political values. For example, Winner discusses the design of bridges over parkways on Long Island, New York, as follows:

> Many of the overpasses are extraordinarily low having as little as nine feet clearance at the curb. Robert Moses, the master builder of roads, parks, bridges, and other public works from the 1920s to the 1970s in New York, had these overpasses built to specifications that would discourage the presence of busses on his parkways. (1980, p. 123)

The overpasses would allow only cars owned by whites from middle and upper classes to freely use the parkways for recreation and commuting, whereas white working-class people and racial minorities traveling by bus would be prevented. In the design of the overpasses, no inputs were solicited by the affected parties, and no ethical concerns were raised. In other words, the issue at stake in Winner's example is therefore that Robert Moses did not consider distributive and procedural knowledge. Wendy Cumming-Potwin and John Currie argued that "if a group cannot participate in planning and decision-making, its interests are unlikely to inform social, political, or financial actions, which can aggravate, rather than reduce inequality" (Cumming-Potwin & Currie, 2013, p. 23). The holistic and democratic mantra "Nothing about us without us" urges engineers to consider the multiple aspects of social justice for engineering solutions, such as oppression, exploitation, marginalization, powerlessness, and cultural imperialism.

To further illustrate the inherent sociotechnical heterogeneity in engineering professionalism and the sociomaterial dynamics pertaining to engineers' design considerations, Winner's example of engineering bridge construction can be elaborated:

> Building a bridge that connects two communities is a technical feat, but success requires social and contextual understanding; designers engage the communities where the bridge will be built, soliciting and addressing concerns, and they explore issues like aesthetics, traffic flow impacts, and how the new bridge will affect the community. (Gravel & Svihla, 2020, s. 6)

The heterogeneity of elements to be included in the design process of bridges are of a social, material, economic, political, technical, environmental, traffic-related, and often, as in the example above, of an aesthetic nature. It seems evident that the sample space of joining together the elements is contingent, as a change in one or several elements is an option that creates alternative solutions. This might happen in a new round of negotiations where new environmental requirements are imposed, uncertainty has arisen, new knowledge becomes available, or budgets are to be curbed. Engineering solutions are diverse, and several alternative solution strategies can be applied. All solutions have benefits and disadvantages, and engineers must continuously deliberate which of them is acceptable, given the constraints of the situation. In weighing up solutions, ethical considerations are either explicitly or implicitly included. Heterogeneity is, to a great extent, a condition for all kinds of professional work. However, the pervasive importance of engineering regarding societies' technological infrastructure necessitates special attention in that it creates technological conditions for other professionals and users by framing their spaces of development.

Even though the educational system generally acknowledges that engineers should exercise ethical judgment, engineering ethics is nevertheless perceived as a minor supplement to an overly instrumental technicist education and is therefore not embedded in engineering epistemology (cf. Murphy et al., 2022). To rectify that situation, it is necessary to rethink engineering education and engineering professionalism by considering a concept of *Bildung* that tackles dichotomous thinking of the social and technical. We turn to John Dewey's philosophy for inspiration.

2.3 Rethinking *Bildung* with Inspiration from John Dewey

It might seem like an odd choice to turn to John Dewey to rethink engineering professionalism through the concept of *Bildung*. Dewey seldom explicitly used the concept of *Bildung* (cf. Pape & Kehrbaum, 2019; Retter, 2012; Wäkevä, 2012), either in his philosophy of education (e.g., Dewey, 2011) or in his discussion of ethics and politics (e.g., Dewey, 1927/2008; Dewey & Tufts, 1932/2008), as he was working in a cultural and intellectual context different from the European context. However, his views on education and growth have strong affinities to the German ideal (Løvlie & Standish, 2003; Siljander et al., 2012). In his early work, Dewey was strongly influenced by German idealism and especially G.W.F. Hegel's philosophy (Shook, 2000). In his mature philosophy, Dewey's ideas on education and growth share some fundamental tenets with a Hegelian conception of *Bildung*, such as the emphasis on the unity of personal and cultural developments, and the importance of education and experience in promoting growth. However, compared with the classical German tradition of *Bildung*, Dewey's development has the advantage of transcending the narrow idealism, essentialism, individualism, and elitism that is often associated with the concept. Instead, he furnishes the concept on the basis of an open-minded secular, democratic, and naturalist pragmatism concerned with inclusiveness, the amelioration of living conditions, and socially responsible action (Rorty, 2021; Wäkevä, 2012) that seem particularly suitable for responding to the challenges confronting engineering professionalism in the Anthropocene era.

To understand the basic tenets of a Deweyan approach, it is important to recognize Dewey's roots in the Darwinian naturalist conception of humanity (Cowles, 2020; Pearce, 2020). According to this conception, human beings are understood as (complex) organisms that survive and evolve by adapting to and constructing livable niches in their environment. However, human adaptiveness must be qualified: it is not a matter of passive isolated organisms that are exposed to and observe environmental challenges and subsequently adapt (or fail to adapt) to them. On the contrary, humans are active social creatures who creatively anticipate environmental challenges by shaping physical, social, cultural, and societal resources to accommodate their lives. The world that humans inhabit is volatile and potentially hostile to their existence, and their continuing survival in these harsh and everchanging environments demands intelligent conduct to "escape from peril" (Dewey, 1929/2008, Chap. 1). The reciprocal relationships between organism and environment, and between humans and the world they inhabit are mediated by transactions that leave neither organisms nor their environment unchanged (Dewey & Bentley, 1948/2008, Chap. 5). Humans do not one-sidedly shape their environment; changes in the environment produced by humans have consequences that feedback and shape their existence. This fundamental ecological insight not only holds sway for humans' transactions with their physical environment but also is important for understanding how selves, social relations, and societies are formatted, cultivated, and shaped, particularly for processes of *Bildung*.

For Dewey, humans are biological creatures who evolve and refine their conduct to thrive and grow. Conduct, that is to say, action, is thus the essential feature, and a *sine qua non* of human nature. It is constitutive of both humans' social lives in societies and how individual selves are shaped. Contrary to the point of departure taken by Cartesian rationalism and classical empiricism and inspired by Hegel's speculative identity thesis (Ng, 2020), Dewey argued that selves and consciousness are not given as isolated, mental, cognitive entities that passively observe and perceive their surroundings (Dewey, 1896/1998). For Dewey, as implied in Hegel's notion of consciousness, selves emerge and are gradually developed as humans actively and purposefully *transact* with each other and their environment to form experiences (Dewey, 1925/2008). Selves are thus thoroughly social and shaped by the habits and interests that play roles in human communities. Dewey and Tufts write:

> It is not too much to say that the key to a correct theory of morality is a recognition of the *essential unity of the self and its acts*, if the latter have any moral significance; while errors in theory arise as soon as the self and acts (and their consequences) are separated from each other, and moral worth is attributed to one more than the other. (Dewey & Tufts, 1932/2008, p. 288)

It is, of course, not the case that actions and selves are literally identical, but the identity is apparent, as the concepts of action and self cannot be conceived independently of each other. They reciprocally constitute each other. Dewey believed that the formations of selves and public culture and society through social action cannot be separated: "Self reveals its nature in what it chooses" (Dewey & Tufts, 1932/2008, p. 287). However, the choices of selves are always collectively structured by habits:

> The essence of habit is an acquired predisposition to *ways* or modes of response, not to acts except as, under special conditions, these express a way of behaving. Habit means special sensitiveness or accessibility to certain classes of stimuli, standing predilections and aversions, rather than bare recurrence of specific acts. It means will. (Dewey, 1922/2007, p. 42)

The unity and consistency of both selves and social culture are constituted by and manifested in habitual actions that are acquired and accumulated in processes of socialization and cultivation. Thus, habits organize goal-directed learners' activities in their efforts to meet and overcome the challenges set by their environment. However, habits must not be conflated with "tradition." Habits do not consult past traditions to determine present action. Instead, they are behavioral dispositions oriented toward dealing with future events, and the future success of dispositions cannot be judged by the standards of a tradition. The habits of organisms are fallible, and their success is only measured in relation to their ability to assist organisms to meet the future challenges of the volatile environment they inhabit.

Even though habits are formative of the self and culture, according to Dewey, they are not all there is to the self. Human selves can also deliberate and choose their course of action, that is, to temper their dispositions and act intelligently (Dewey, 1922/2007, pp. 172–180). When confronted with problematic situations that interrupt habitual action, humans have the ability to hesitate and engage in reflection on strategies to find new and more successful ways to achieve their goals. This ability to pause habitual action, reflect on the unsuccessfulness of previous attempts to solve problems, and intelligently suggest new avenues for future action that potentially may deal with the problematic situation characterizes intelligent human thinking (Dewey, 1909) and, when refined and systematized, scientific inquiry and reasoning (Dewey, 1938/2008). For Dewey, the human ability to learn must be understood in terms of intelligent inquiry that enables the transformation of conduct and habits. Like Humboldt's conception of *Bildung*, Dewey's conception of learning is thus characterized as a transaction between humans and the world that enables humans to acquire knowledge about the world they inhabit and grow. However, Dewey draws an important distinction between knowledge and coming to know. Unlike the representational conception of knowledge in the German idealist tradition, which understands knowledge as the mind's truthful representation of the world, Dewey does not conceive knowledge as a fixed, objective entity but rather as a dynamic, transactional process that is shaped by the needs and interests of individuals and communities. For Dewey, knowledge is the fallible and preliminary *product* of people coming to know, that is, organisms' purposive transactions with their environment. Humans use this product as a resource and instrument for shaping new habits that will allow them to act successfully in the future.

Besides construing knowledge in instrumental and performative terms, a Deweyan conception of *Bildung* will also diverge significantly from classical conceptions in other ways. Whereas the ideals of *Bildung* held in the classical German tradition strives toward the perfection of intellect, sensibility, and judgment toward *absolute and fixed standards* of autonomy and moral worth, a Deweyan rendering of learning does not posit absolute standards. The telos of learning, moral worth, and valuation may vary and alter from person to person, from one situation to another, and over the course of people's lives. The telos of learning is not set as a transcendental goal but as an immanent normative standard of accountability toward self-development, that is, as the adaptation of the human organism to thrive in its environment:

> A human being is held accountable in order that he may learn; in order that he may learn not theoretically and academically but in such a way as to modify and—to some extent—remake his prior self. The question of whether he might when he acted have acted differently from the way in which he did act is irrelevant. The question is whether he is capable of acting differently next time; the practical importance of effecting changes in human character is what makes responsibility important. [...] With every increase of capacity to learn, there develops a larger degree of accountability. [...] The fact that each act tends to *form*, through habit, a self which will perform a certain kind of acts, is the foundation, theoretically and practically of responsibility. We cannot undo the past; we can affect the future. (Dewey & Tufts, 1932/2008, p. 304)

Dewey's theory of experiential learning is often misinterpreted to hold that learning is caused by experience. However, learning only comes about when the self *reflects* on experience and accommodates the present situation with goals set to alter it. Thus, learning becomes the vehicle for the transformation of self and its environment; that is, it is the vehicle of the reciprocal process of *Bildung* of the self and society.

Another misconception of a Deweyan construal of *Bildung* holds that by negating absolute and fixed standards, we embark on a postmodern ateleological construal of *Bildung* that resembles "the eternal building site [where] we can admit that there are no goals of *Bildung*" (Reichenbach, 2003, p. 98). However, Dewey's negation of absolute and fixed goals for learning does not claim that there are no goals and, significantly, that all goals have equal moral worth. People's learning is guided by different goals and objectives. However, not positing absolute and unchangeable moral principles need not imply embarking on relativism. Dewey believed that ethical principles are grounded in our experiences and that they evolve over time as we learn from critically reflecting on our experiences and the consequences of our actions. Dewey's moral philosophy is contextual, neither absolutist nor relativistic. Moral principles must be continually reevaluated and renegotiated in response to changing circumstances in the environment so that humans can adapt and thrive.

> The kind of self which is formed through action which is faithful to relations with others will be a fuller and broader self than one which is cultivated in isolation from or in opposition to the purposes and needs of others. In contrast, the kind of self which results from generous breadth of interest may be said alone to constitute a development and fulfillment of self, whilst the other way of life stunts and starves selfhood by cutting it off from the connections necessary to its growth. (Dewey & Tufts, 1932/2008, p. 302)

Reconstruing *Bildung* from a Deweyan naturalist perspective does not fix the telos of learning except by gesturing to the fact that humans, as organisms, must grow to exist: "We set up this and that end to be reached, but *the* end is growth itself" (Dewey & Tufts, 1932/2008, p. 306). Growth becomes the penultimate telos of *Bildung*, but the road that leads to *Bildung* is not fixed. As learners, educators, citizens, and so on, we must conduct our actions to stimulate the growth of selves and society that can lead to further growth. This means that growth, both personal and societal, must be sustainable. It must "set [...] thought and desire upon the *process* of evolution instead of upon some ulterior goal [...]. It is the next step which lies within our power" (ibid.). Growth must be flexible and driven by a telos that enables the continuation of organisms' growth. Having reconstructed the concept of *Bildung*, we can now proceed to rethink engineering professionalism in ways that makes it able to meet future challenges.

2.4 Rethinking Engineering Professionalism Through the Concept of *Bildung*

Recognizing that the telos of a Deweyan reconstruction of *Bildung* is growth in and for itself prompts us to reconsider the telos of engineering professionalism. Buch (2022) argued that engineering must be recovered to meet the challenges of the future and that this recuperation must start with a reexamination of the telos of engineering, which is the very purpose and "in-order-to" dimension of the profession. The *Cambridge Dictionary* defines *engineering* as the "use of scientific principles to design and build machines, structures, and other items, including bridges, tunnels, roads, vehicles and buildings." However, this vernacular and functional definition fails to address the reasons, objectives, and purposes of engineering activities. Instead, Buch suggested a more encompassing definition that incorporates a telos for the profession in terms that "envision […] engineering as a goal-oriented activity of designing and making artifacts and systems to exploit natural phenomena for human ends, mostly (but not necessarily always) by using scientific principles, techniques, and results" (Buch, 2022, p. 363). This suggested alternative definition, while still functional and instrumental, includes a telos for engineering, namely "exploit[ing] natural phenomena for human ends."

It should be noted that while the alternative definition of "engineering" installs a telos, it does not posit an absolute normative value standard for "human ends." Furthermore, recognizing that human ends are plural (and often conflicting) and might change as new challenges and problems arise in the environment of humans, the characterization of the objectives of engineering must be determined relationally and not absolutely. Our Deweyan reconstruction of the concept of *Bildung* as the relational growth of humans in relation to their environment provides a useful new conceptual platform for rethinking engineering professionalism and clarifying how a telos of engineering should be understood. Notably, rethinking implies the following:

> Instead of conceptualizing engineering as an applied science that takes its cues from science and, to a much lesser extent, from [the social sciences and humanities], engineering must be reconstructed as a practical, purposive human activity that instrumentally uses specialized means and techniques from the sciences and [the social sciences and humanities] to solve the relevant and pressing problems that humans encounter in concrete situations in their daily lives. (Buch, 2022, p. 364)

While the Deweyan reconstruction of *Bildung* is generic and not restricted to any human endeavors in particular, its appropriateness to inform engineering professionalism as sociotechnical problem-solving must be stressed. The purposiveness of engineering endeavors must be split from neither the technical knowledge and epistemic values involved in intervening with natural phenomena nor the social and cultural ends that guide the activity. The technical and social dimensions of engineering are entangled in the means-end relationships that direct activities in the profession. Only acknowledging the technical dimension results in what Theodor W. Adorno aptly diagnosed as *Halbbildung*, or the fragmentation, commodification, and assimilation of objects into pre-given subjective schemes of interpretation of reality that prevent the generation of experience (Adorno, 1959/2006). Including and recognizing the social, cultural, and societal dimensions of engineering is pertinent for the ability to integrate and expand experience in ways that meet "human ends." Winner's case of bridge constructions in New York illustrates how technical problem-solving (the construction of bridges) unavoidably presumes a social and cultural problem framing (whose and what problems are addressed) that determines how "human ends" are interpreted, specified, and enacted in particular social, cultural, and societal situations.

Considering a Deweyan conception of *Bildung* helps us realize that engineering professionalism becomes *Halbbildung* if its scope is reduced to the "application" of "pure" or "neutral" scientific principles. The enactment of engineering practices has "politics" and is vested in moral frameworks that make engineering problem-solving accountable to human ends. A Deweyan conception of *Bildung* makes it pertinent to consider how engineering might serve human ends to allow personal and social growth and to reflect on how engineering activities avoid focusing on short-term growth that eventually obstructs flourishment and further growth in a long-term perspective. When growth is set as the penultimate goal of the profession, technical quick fixes and unreflected problem-solving that do not take the social embeddedness of technical challenges into account must be dismissed by engineering professionals. Dewey argued that the only goal of growth is growth itself, implying that growth must be sustainable to serve human ends.

2.5 Barriers and Obstacles

As we have argued, Dewey's conception of *Bildung* has the potential to deal with hypercomplex challenges and crises that societies, nature, and humankind face in a globalized world. Thus, the Deweyan framework helps acknowledge that global warming, pandemics, decreasing biodiversity, economic and social polarization, racism, information, and cyberwars and cybercrimes are complex problems that cannot be solved by narrowly envisioned technical fixes. They are problems that can only find solutions from a long-term perspective that carefully considers sociotechnical means and ends to achieve sustainable growth. An isolated technical expertise sustained by narrow disciplinary boundaries is therefore no longer sufficient. Thus, Robert Frodeman (2014) pointed to the need to think technology and society together through interdisciplinary, de-disciplinary, and problem-driven education and research.

Moreover, Edmond Byrne and colleagues in engineering education and social sciences argued that transdisciplinary research with SSH is a necessary partnership in addressing contemporary (un)sustainability challenges (Byrne et al., 2022). This also means that engineering should be identified as a truly heterogeneous sociotechnical enterprise associated with an engineering education redesigned to fulfill that purpose. Such a venture necessitates thoroughgoing conversations between engineers, social scientists, and scholars from the humanities in developing educational programs that can shape the competence profile of future generations of engineers. Barriers and obstacles to such conversations are, unfortunately, abundant. While they can be identified on many levels, it is useful to shorthand them around a sociocultural and institutional axis.

Regarding the first axis, the sociocultural, strong evidence shows that it serves as a major impediment for change in engineering education (cf. Jensen & Cross, 2021). Briefly put, engineering culture, according to these authors, can be described as a masculine, weed-out, and stress culture and as a culture of suffering and shared hardship often tied to mathematical ability, which is less welcoming to women and students of color. These characteristics of engineering culture seem to be deeply rooted and widely dispersed across national cultures (Kunda, 2009; Meiksin et al., 2020; Mellström, 1995). This is aggravated by the following fact:

> Engineering disciplines described as less welcoming are also the largest producers of engineering degrees, despite women and students of color remained underrepresented. (Roy, 2018)

Determined as a disciplinary culture, engineering identity is typically portrayed as singular and homogeneous. However, owing to identity variations, it is also more complex. Thus, engineering identities may vary from heterogeneous identities to identities rooted in engineering science, with the former still in the making. Where the former may be considered as enhancing the inclusion of *Bildung* in engineering degree programs, the latter may be a major impediment.

Engineering ethics as an academic subject is a newcomer in engineering curricula. As an academic field of study, it was established in the 1970s. By tradition, engineering was considered a vehicle for social and economic progress; therefore, morally good and ethical education was deemed unnecessary. As a result, the development of the subject has been slow in the making (Martin et al., 2021). Furthermore, a major body of literature testifies to the challenges and impediments of including the subject in engineering curricula. Habitually, ethics courses are taught by SSH scholars in parallel with and, to a large extent, isolated from the technical parts of engineering curricula. In the absolute minimal version, engineering ethics is reduced to a simple code of conduct in a less than welcoming environment (Murphy et al., 2022). We admit that positive exceptions exist. Hence, whether engineering epistemology seamlessly amalgamates the technical and sociocultural dimensions of engineering is yet to be seen.

These observations are closely related to an epistemology that has inherited the dichotomized tropes of Western philosophy: subject/object, mind/body, social/technical, etc. (cf. Dewey, 1947/2023). Briefly, a major impediment to changing engineering into a heterogeneous sociotechnical enterprise is that engineering epistemology is still largely molded upon a *technical core/non-technical periphery* dichotomy that constitutes the prevalent competence profile of engineers, with bearings on engineering identities, although as noted earlier, heterogeneous engineering identities at the same time do exist and flourish. Accreditation criteria at the policy level have been highlighted to potentially enhance the role of ethics through the Washington Accord. Nevertheless, as reported by Martin et al. (2021), there are doubts as to whether the pressure from accreditation criteria can boost engineering educators' commitment to deeper curricular change.

Barriers and obstacles are also apparent at the institutional level. Educational institutions and work organizations in corporate companies are enacted by sociocultural practices that impose structurally enforced boundaries, rules, and textures of learning that reflect the prevailing epistemological dichotomies (Gherardi, 2009). Stephen Kemmis et al. (2014) coined the term "ecology of practices" to describe the interconnectedness and dynamic relationship between social practices and institutional settings. In an ecology of practices, various social practices such as teaching, learning, assessment, curriculum design, and classroom management are intertwined and mutually influence each other. This means that changes in one practice can have an impact on the others, which leads to a ripple effect throughout the entire educational system. Conversely, dominant, and resilient social practices are likely to keep other practices in check and prevent overall change on an institutional level. Studies have shown how holistic-minded approaches to engineering are hampered in ecologies of practices, making it difficult to "make room" for fostering and enacting sociotechnical engineering approaches (e.g., Buch, 2016; Petersen & Buch, 2016).

2.6 Conclusion

Evoking the German concept of *Bildung* that some would readily declare as a non-operational and outdated concept and rethinking it along Dewey's naturalist and democratic philosophy might thus seem a cumbersome and odd road to embark on when considering the challenges that contemporary engineering professionalism faces. It might even seem an extremely difficult road to take, as the obstacles are manifold and clearly in view. Thus, why not make use of a more recent and specialized conceptual framework that promises immediate results and has already gained traction in engineering programs? The reason for not doing so concerns the nature of the contemporary challenges and obstacles that engineers face when trying to solve them. The major contemporary challenges faced in the Anthropocene era are, to a large extent, problems that humans and industrial societies have brought upon themselves.

The crafting of the fossil economy and the associated infrastructure of contemporary societies are, for better or worse, examples of the achievements of engineering professionalism. It has brought wealth and prosperity but only to some societies and social groups and at the cost of ecological disasters and global warming. The lesson learned is that most engineering problems are extremely complex and sociotechnical in nature and that finding sustainable solutions therefore requires that engineers consider the problems along a broad and future-oriented horizon. Engineers shape nature *and* society, and the effects of engineering solutions frequently have dramatic consequences for humans' ability to learn and grow and ultimately to survive, as they impact humans' environment.

Thus, engineering professionalism must be rethought out to meet the challenges of the future and become sustainable. It is not only the technical solutions of engineers but also the mindset of the engineers themselves, along with the institutions that educate them for the future, that must be sustainable. Dewey's rethinking of *Bildung* as a framework that insists on critical thinking, learning, and growth extends the ecological perspective of nature to culture. Redefining the telos of engineering as a sociotechnical problem-solving endeavor that strives for sustainable growth sets a goal not only for engineering professionalism but also for engineering education. Engineering education must provide future engineers with technical skills and, more importantly, habits for sustainable problem-solving. Such habits are not fixed at one time for all. They must be flexible enough to accommodate the nature of the everchanging situations and contexts of engineering work. Dewey saw the primary task of education as the stimulation of learning and nourishing a natural inquisitiveness and an exploratory attitude:

> The native and unspoiled attitude of childhood, marked by ardent curiosity, fertile imagination, and love of experimental inquiry, is near, very near, to the attitude of the scientific mind. (Dewey, 1909, preface)

Dewey called such attitudes and habits "intelligence." They embody a reverent awe toward possibilities and the unfinished and a moral accountability that is stimulated by the ability to imagine the consequences of one's actions and the suffering of others.

Reconstructing engineering professionalism along the insights provided by a Deweyan concept of *Bildung* embodies the basic idea of "Build Back Better," that is, the idea to use the opportunity of reconstructing efforts to address *underlying* issues that may have contributed to the crisis in the first place and create a more inclusive, equitable, and sustainable future. Employing the concept of *Bildung* provides the realization that shaping a sustainable future and developing engineering selves with a habit of engaging in sustainable problem solving are connected activities.

References

Adorno, T. W. (1959/2006). *Theorie der Halbbildung*. Suhrkamp.

Alves, A. (2019). The German tradition of self-cultivation (*Bildung*) and its historical meaning. Available at http://creativecommons.org/licenses/by/4.0

Buch, A. (2016). Ideas of holistic engineering meet engineering work practices. *Engineering Studies, 8*(2), 140–161.
Buch, A. (2022). The need for recovery of engineering. In S. H. Christensen, A. Buch, E. Conlon, C. Didier, C. Mitcham, & M. Murphy (Eds.), *Engineering, social sciences, and the humanities: Have their conversations come of age?* Springer Nature.
Byrne, E., Keohane, K., Revez, A., Boyle, E., McGookin, C., Dunphy, N., O'Neill, C., Harris, C., Hughes, I., Sage, C., Barry, J., Gallachóir, B. Ó., & Mullally, G. (2022). Engineering with social sciences and humanities in facing contemporary (un)sustainabiliy changes. In S. H. Christensen, A. Buch, E. Conlon, C. Didier, C. Mitcham, & M. Murphy (Eds.), *Engineering, social sciences, and the humanities: Have their conversations come of age?* Springer Nature.
Christensen, S. H. (2019). Costs and benefits of commercializing teaching, research, and service in the American Corporatized University. In S. H. Christensen, B. Delahousse, C. Didier, M. Meganck, & M. Murphy (Eds.), *The engineering-business nexus: Symbiosis, tension and co-evolution*. Springer Nature.
Christopherson, S., Gertler, M., & Gray, M. (2014). Universities in crisis. *Cambridge Journal of Regions, Economy and Society, 7*(2), 209–215.
Cowles, H. (2020). *The scientific method. An evolution of thinking from Darwin to Dewey*. Harvard University Press.
Cumming-Potwin, W., & Currie, J. (2013). Towards new literacies and social justice for engineering education. *International Journal of Engineering, Social Justice, and Peace., 2*(1), 21–37.
Dewey, J. (1896/1998). The reflex arc concept in psychology. In L. A. Hickman & T. M. Alexander (Eds.), *The essential Dewey* (Ethics, logic, psychology) (Vol. 2). Indiana University Press.
Dewey, J. (1909). *How we think*. D.C. Heath 6 Company.
Dewey, J. (1922/2007). *Human nature and conduct. An introduction to social psychology*. Cosimo.
Dewey, J. (1925/2008). In A. B. Carbondale (Ed.), *Experience and nature, the later works* (Vol. 1). South Illinois University Press.
Dewey, J. (1927/2008). In A. B. Carbondale (Ed.), *The public and its problems, the later works* (Vol. 2). South Illinois University Press.
Dewey, J. (1929/2008). In A. B. Carbondale (Ed.), *The quest for certainty, the later works* (Vol. 4). South Illinois University Press.
Dewey, J. (1938/2008). In A. B. Carbondale (Ed.), *Logic, the later works* (Vol. 12). South Illinois University Press.
Dewey, J. (1947/2023). *Unmodern philosophy and modern philosophy*. Southern Illinois University Press.
Dewey, J. (2011). *Democracy and education*. Simon & Brown.
Dewey, J., & Bentley, A. (1948/2008). In A. B. Carbondale (Ed.), *Knowing and the known, the later works* (Vol. 16). South Illinois University Press.
Dewey, J., & Tufts, J. H. (1932/2008). In A. B. Carbondale (Ed.), *Ethics, the later works* (Vol. 7). South Illinois University Press.
Etzkowitz, H. (2008). *The triple helix. University-industry-government innovation in action*. Routledge.
Frodeman, R. (2014). *Sustainable knowledge: A theory of interdisciplinarity*. Palgrave Macmillan.
Gadamer, H.-G. (1969/2013). *Truth and method*. Bloomsbury Academic.
Gherardi, S. (2009). *Organizational knowledge. The texture of learning*. Blackwell Publishing.
Gibbons, M., Limoges, C., Nowotny, H., Schwartzman, S., Scott, P., & Throw, M. (1994). *The new production of knowledge. The dynamics of science an research in contemporary society*. SAGE Publications.
Gravel, B. E., & Svihla, V. (2020). Fostering heterogeneous engineering through whole-class design work. *Journal of the Learning Sciences, 30*, 279.
Hammershøj, L. G. (2017). *Dannelse i Uddannelsessystemet*. Hans Reitsels Forlag.
Hansen, T. B., Nielsen, K. H., Troelsen, R. P., & Winther, E. (2000). *Naturvidenskab, Dannelse og Kompetence*. Aalborg Universitetsforlag.

Humboldt, W. v. (2000). Theory of *Bildung*. In I. Westbury, S. Hopmann, & K. Riquarts (Eds.), *Teaching as a reflective practice* (The German didaktik tradition). Routledge.

Humboldt, W. v. (2017). *Wilhelm von Humboldt: Schriften zur Bildung*. Reclam.

Jensen, K. J., & Cross, K. J. (2021). Engineering stess culture: Relationships among mental health, engineering identity, and sense of inclusion. *Journal of Engineering Education, 110*, 371–392. Wileyonlinelibrary.com/journal/jee

Kemmis, S., Wilkinson, J., Edwards-Groves, C., Hardy, I., Grootenboer, P. & Bristol, L. (2014). *Changing Practices, Changing Education*. Springer: Singapore.

Konrad, F.-M. (2012). Wilhelm von Humboldt's contribution to a theory og *Bildung*. In P. Siljander, A. Kivelä, & A. Sutinen (Eds.), *Connections and controversies between continental educational thinking and American pragmatism* (pp. 107–124). Sence Publishers Rotterdam.

Kunda, G. (2009). *Engineering culture. Control and commitment in a high-tech corporation*. Temple University Press.

Løvlie, L., & Standish, P. (2003). Introduction: *Bildung* and the idea of a liberal education. In L. Løvlie, K. P. Mortensen, & S. E. Nordenbo (Eds.), *Educating humanity: Bildung in postmodernity*. Blackwell Publishing.

MacKenzie, D., & Wajcman, J. (Eds.). (2005). *The social shaping of technology* (2nd ed.). Open University Press.

Martin, D. A., Conlon, E., & Bowe, B. (2021). A multi-level review of engineering ethics education: Towards a socio-technical orientation of engineering education for ethics. *Science and Engineering Ethics, 27*, 27–60.

Meiksin, P., Leyne, P., Beddoes, K., & Deters, J. (2020). Women in engineering: A review of the 2019 literature. *Magazine of the Society of Women Engineers, 69*(2), 4–38. https://magazine.swe.org/lit-review-22/

Mellström, U. (1995). *Engineering lives. Technology, time and space in a male-centered world*. Linköping University Press.

Miller, A. (2020). The promise of *Bildung- or* 'a world of one's own. *Journal of Philosophy of Education, 55*(2), 334.

Murphy, M., Christensen, S. H., & Conlon, E. (2022). Social justice at an Irish practice-based university: In or out of place? In S. H. Christensen, A. Buch, E. Conlon, C. Didier, C. Mitcham, & M. Murphy (Eds.), *Engineering, social sciences, and the humanities: Have their conversations come of age?* Springer Nature.

Ng, K. (2020). *Hegel's concept of life. Self-conciousness, freedom, logic*. Oxford University Press.

Pape, H., & Kehrbaum, T. (2019). *John Dewey: Über Bildung, Gewerkschaften und die demokratische Lebensform*. Study der Hans-Böckler-Stiftung, No. 421.

Pearce, T. (2020). *Pragmatism's evolution. Organism and environment in American philosophy*. Universituy of Chicago Press.

Petersen, R. P., & Buch, A. (2016). Making room in engineering design practices. *Engineering Studies, 8*(2), 93–115.

Reichenbach, R. (2003). On irritation and transformation: A-teleological *Bildung* and its significance for the democratic form of living. *Journal of Philosophy of Education, 36*, 3.

Retter, H. (2012). Dewey's progressive education, experience and instrumental pragmatism with particular reference to the concept of *Bildung*. In P. Siljander, A. Kivelä, & A. Sutinen (Eds.), *Connections and controversies between continental educational thinking and american pragmatism*. Sence Publishers Rotterdam.

Rohstock, A. (2012). Some things never change: The invention of Humboldt in western higher education systems. In P. Siljander, A. Kivelä, & A. Sutinen (Eds.), *Connections and controversies between continental educational thinking and American pragmatism* (pp. 165–182). Sence Publishers Rotterdam.

Rorty, R. (2021). *Pragmatism as anti-authoritarianism*. The Belknap Press of Harvard University Press.

Roy, J. (2018). *Engineering by the numbers*. American Society for Engineering Education. Retrieved from Google Scholar 6. March 2023.

Shook, J. (2000). *Dewey's empirical theory of knowledge and reality*. Vanderbilt University Press.
Siljander, P., Kivelä, A., & Sutinen, A. (Eds.). (2012). *Connections and controversies between continental educational thinking and American pragmatism*. Sence Publishers Rotterdam.
Sjöberg, S. (2005). *Naturfag som Almendannelse*. Klim.
Sjöström, J., Frerichs, N., Zuin, V. G., & Eilks, I. (2017). Use of the concept of *Bildung* in the international science education literature, its potential, and implications for teaching and learning. *Studies in Science Education, 53*(2), 165–192.
Wäkevä, L. (2012). Experiencing growth as a natural phenomenon: John Dewey's philosophy and the *Bildung* tradition. In P. Siljander, A. Kivelä, & A. Sutinen (Eds.), *Theories of Bildung and growth* (pp. 261–279). Sense Publishers.
Winner, L. (1980). Do artefacts have politics? In *Daedalos* (Modern technology: Problem or opportunity?) (Vol. 109(1), pp. 121–136). The MIT Press.

Anders Buch Docent at VIA University College, Denmark and affiliated researcher at Jönköping University, Sweden. He holds an MA in philosophy from University of Copenhagen and a BEd and PhD in education from Roskilde University. His empirical research area is focused on professions, professionalism, and technological expert cultures, and his theoretical approach is primarily inspired by Science & Technology Studies, Practice Theory, and classical American Pragmatism. He has published articles and books on knowledge, learning, education, professionalism, and the professional development of engineers. Since 2016 he has servedas chief editor of Nordic Journal for Working Life Studies.

Steen Hyldgaard Christensen († 23rd June 2024). Independent researcher. He held an MA in Danish Language and Literature and the History of Ideas from Aarhus University, and PhD in research on engineering education from Aalborg University. Until 2014 senior lecturer at Aarhus University, Herning. From 2014 until May 2023, he was an adjunct associate professor at Aalborg University. He was initiator, coordinator, lead editor, and co-author of several edited volumes on engineering education, and epistemology. Five of them are published in the Springer Nature (POET) series. He has published articles on engineering epistemology, culture, and higher education. His research interests included the dynamics in higher education restructuring, and research initiatives to broaden engineering education.

Chapter 3
Bildung Under Conditions of Engineered Destruction

Carl Mitcham

Abstract What might it mean to aspire to the realization of a specifically engineering *Bildung* under conditions in which engineering has become a generative source for social-political and planetary disorder? Taking off from another effort to think engineering professionalism in engineering today, this chapter revisits the shift from pre-Enlightenment to Enlightenment concepts of *Bildung*, some of the boundary assumptions behind the notion of a possible engineering *Bildung*, and the mythopoeic engineer scientist J. Robert Oppenheimer as an exemplar of apocalyptic engineering *Bildung*.

Keywords Anders Buch and Steen Hyldgaard Christensen *Bildung* in engineering · Jennifer Herdt · J. Robert Oppenheimer

3.1 Introduction

> Now I am become Death, the destroyer of worlds.
> —J. Robert Oppenheimer, Alamogordo, New Mexico, July 16, 1945

In their "Rethinking Engineering Professionalism through the Concept of *Bildung*," Anders Buch and Steen Hyldgaard Christensen begin with anchoring quotations from two German philosophers: Wilhelm von Humboldt (1767–1835) and Hans-Geog Gadamer (1900–2002). For von Humboldt, *Bildung* aims at "linking the self to the world to achieve the most general, most animated, and most unrestrained interplay." For Gadamer, it aims at making one's profession "wholly one's own …

C. Mitcham (✉)
Colorado School of Mines, Golden, CO, USA

Beijing, China
e-mail: cmitcham@mines.edu

A. Buch, S. H. Christensen (eds.), *Bildung for Engineering Education and Practice*, Philosophy of Engineering and Technology 49,
https://doi.org/10.1007/978-3-031-86581-7_3

wholly one's concern" (Buch & Christensen, 2025, p. 20). The first assumes that the world is well ordered, encompassing, and fundamentally good; the second that one's profession is engaged in practices to make things better. What does *Bildung* become when the world is disordered, and professional practice is contributing to that disorder—not to say its destruction?

This is not exactly Buch and Christensen's question, but it's related, and I want to consider it from three angles that complement and deepen their analysis. One concerns the broader meaning of and debate about *Bildung* itself, another some Buch and Christensen boundary assumptions, and, finally, prospects for what I will call apocalyptic engineering *Bildung*.

3.2 Forming Humanity: The German *Bildung* Tradition

In *Forming Humanity: Redeeming the German Bildung Tradition*, Jennifer Herdt (2019) deepens and problematizes the von Humbolt and Gadamer anchorings. While recognizing the central, Enlightenment educational ideal of *Bildung* as the fully formed (from *bilden*, "to form, to shape, to construct") capaciously humanized individual developed within a community of seventeenth century German philosophers (Kant, Herder, Goethe, Schiller, von Humboldt, and Hegel), she traces its roots back to Greek *paideia* and late medieval-early modern Christian pietism. Pursuit of a secular cosmopolitan, interdisciplinarily, ethically, and aesthetically educated individualized humanity was preceded by a reformed vision of Christian personhood that was as critical of Catholic ritualism as of Enlightenment philosophy.

Early Protestant theology preached persons as sinful creatures who were nevertheless created in the image (*Bild*) of God and as such called on to search within for the divine model (*Vorbild*).

> The concept of *Bildung* initially took shape in a specifically theological context. Meister Eckhart and other Rhineland mystics, employing for the first time the German vernacular to discuss such matters, took up the term *Bildung*, and its many variants formed through the addition of a wealth of nominal and verbal prefixes, to speak of divine formative and re-formative agency: of God's formation of human being in the divine image, of the fall as the destruction or defacing of that image, and of the re-formation of fallen humanity after the image of God in Christ. (Herdt, 2019, p. 7)

Herdt calls attention especially to how introspective German pietism prepared the way for the secular ideal manifested in the *Bildungsroman* and how Jacob Böhme (1575–1624) gave the inner *Bildung* experience cosmic significance. Drawing on the hermetic tradition found in Paracelus, Böhme envisioned "processes unfolding within the Godhead [as corresponding to] processes happening in nature and individual creatures" (p. 65). For Böhme the faculty for recognizing the correspondence is imagination,

> the power of the mind's eye, to see with closed eyes and in the dark, to travel virtually while rooted on the spot … a mysterious, even magical power … for creation rather than destruction…. By virtue of possessing imagination, *Einbildung*, human beings are specially con-

> ceived in the image of God…. Böhme goes so far as to say that "everything arises through imagination." (p. 66)

From this perspective *Bildung* becomes more than the perfection of humanity; it is the perfection of a wholisticly conceived, human including cosmos. History is a process of divine self-realization that culminates in human *Bildung*. The pivotal transformation of this theological-ethical *Bildung* into the strictly secular ethical-aesthetic was Immanuel Kant's definition of Enlightenment as

> the human emergence from a self-incurred immaturity (*Unmündigkeit*). Immaturity is the inability to use one's own understanding without another's guidance. It is self-imposed when its cause lies not in lack of understanding but in indecision and lack of courage to use your *own* mind without another's guidance. *Sapere aude!* "Have the courage to use your own understanding!" is thus the Enlightenment motto. ("Answering the Question: What Is Enlightenment?" 1784)

The Enlightenment challenge to human beings to courageously declare their autonomy and become rational self-creators is to be realized through the modern scientific expulsion of gods from human theory and the engineering enhancement of human practice—although the engineering element of self-willed pursuit of human maturity is customarily slighted and subsumed within science. The word "science" has more prestige than "engineering" or "technology," even though much of that prestige depends on its incorporation into engineering.

The god-deprived, engineeringly constructed world is thought in romantic philosophy in terms of poetics. For Percy Bysshe Shelley, "Poets are the unacknowledged legislators of the world" ("The Defense of Poetry," 1821). In the wake of Kant there emerged with Friedrich Schiller and von Humboldt a program of ethic-aesthetic formation—a religion of art (*Kunstreligion*)—as replacement for Christianity. What was remarkably not imagined in this religion of art or considered in Herdt's historical study is engineering as either *Bildung* or poetry. But it was engineers and engineering, in the hands of newly liberated capitalists, that gave us the Industrial Revolution—a far more consequential transformation than either poetry or art.

One exception is the German electrical engineer philosopher Friedrich Dessauer, who argued in the early twentieth century for recognizing engineering practice as "the greatest earthly experience of mortals" (Dessauer, 1972 [1927], p. 331), implicitly framing it as the ultimate *Bildung*. Something similar might be attributed to the American civil engineer Samuel Florman and his argument for "the existential pleasures of engineering" (Florman, 1976).

Across the 18th and 19th and into the twentieth century the *Bildung* ideal exercised continuing influence in the German speaking world. Since the mid to late twentieth century, however, the tradition has been increasingly subject to widespread criticisms as bourgeois or elitist. Somewhat idiosyncratically, Herdt focuses on two particular versions of these criticisms: One is Protestant theologian Karl Barth's argument that the *Bildung* tradition is vitiated by an excessive optimism regarding the powers of human agency that understands the human as "the measure of all things" and leaves no room for divine revelation, providence, or grace. A second is a post-colonial theory criticism that professional "*Bildung* is bound up with racism, justification of slavery, colonialism, imperialism, and the origins of National Socialism" (p. 13). Needless to add, these are two criticisms that can be leveled against engineering as well.

3.3 The Buch-Christensen Argument

Taking engineering as a specific professional practice in question, Buch and Christensen reimagine a technical *Bildung* that would address "technological challenges and broader socially pertinent problems … such as global warming, pandemics, decreasing biodiversity, increasing economic and social polarization, racism, cyberwars and cybercrimes" (Buch & Christensen, 2025, p. 21). Engineering *Bildung* so conceived implicates both technical education curricula and "political philosophy of engineering." This is among the few reflections on engineering to so explicitly link engineering pedagogy and political philosophy. Much more common are arguments for engineering ethics, but Buch and Christensen clearly recognize that "ethics is not enough" (Mitcham, 2015).

For Buch and Christensen, engineering is a "heterogeneous sociotechnical enterprise" that contributes "to creating not only the material and technological infrastructure of societies but also the professional, social, and cultural spaces we inhabit" (Buch & Christensen, 2025, p. 22). Engineering educational programs, however, seldom pay more than lip service to ethical or political responsibilities that such broad influences entail. Indeed, for more than two decades, Steen Christensen has subjected this lacuna to critical interrogation along with proposals for melioration in a series of collectively edited volumes that trace back two decades (see Christensen et al., 2007). Neither engineering curricula nor professional engineering societies foreground the Janus face of engineering knowledge and practice. Silently correcting Dessauer and Florman, Buch and Christensen note that the engineering "of the fossil economy and the associated infrastructure of contemporary societies are, for better or worse, examples of the achievements of engineering professionalism [that have] brought wealth and prosperity but only to some societies and social groups and at the cost of ecological disasters and global warming" (Buch & Christensen, 2025, p. 31).

To infuse this correction into engineering education in a way that goes beyond any narrow focus on individualist engineering ethics, Buch and Christensen reach out to John Dewey and his bio-social evolutionary humanism. As they note, Dewey didn't make much use of the term *Bildung*, but his broad vision of education exhibits both similarities and differences from the German ideal.

As for similarities, with Dewey education is simply an extension on the level of culture of a process that takes place at the biological level when individual organisms are born and mature and when species evolve through interaction with their environment: "what nutrition and reproduction are to physiological life, education is to social life" (Dewey, 1944, p. 9). Ontologically, the "primary condition of growth is immaturity," which "means something positive, not a mere void or lack," but "the possibility of growth … the ability to develop" (pp. 41–42). "Since growth is the characteristic of life, education is all one with growing; it has no end beyond itself [and is to be valued by] the extent in which it creates a desire for continued growth and supplies means for making the desire effective in fact" (ibid. p. 53).

One difference argued by Buch and Christensen is that Dewey's ideal "has the advantage of transcending the narrow idealism, essentialism, individualism, and elitism that is often associated with [*Bildung*]." In contrast with German aspirations to create an Enlightenment version of Greek and Renaissance culture, Dewey aspires to an "open-minded secular, democratic, and naturalist pragmatism concerned with inclusiveness, the amelioration of living conditions, and socially responsible action" (Buch & Christensen, 2025, p. 24). Reconstructing *Bildung* along these lines "does not fix the telos of learning except by gesturing to the fact that humans, as organisms, must grow to exist: 'We set up this and that end to be reached, but *the* end is growth itself'" (Buch & Christensen, 2025, p. 27).

The urge to growth, as it were, comes as a kind of push from below to survive rather than a pull from above to reach an ideal. In this regard pragmatist growth would seem to privilege technology as a process, that is, engineering. It seems a peculiarly apt ideology for a techno-capitalist society such as ours that knows no limits to growth. Evolutionary processes disturb existing orders while creating new ones. Of course, Dewey was a critic of capitalism in the name of social justice, but not in the name of harmony with or contributing to a cosmic order. Nor does it contribute anything distinctive to the cosmological process. It perfects only itself; the larger processes of which it is a part are more or less meaningless. Meaning comes from the human.

> As the geological record shows, not merely individuals but also species die out [and] as some species die out, forms better adapted to utilize the obstacles against which they struggled in vain come into being. Continuity of life means continual readaptation of the environment to the needs of living organisms. (Dewey, 1944, p. 2)

Buch and Christensen recognize at least one aspect of this deficiency in the pragmatist philosophy of education, when they attempt to apply it to rethinking engineering professionalism. Somehow the "contingently imposed human end that doesn't undermine "*the* end [of] growth itself," will need to be reconfigured for engineering from what it was at its origins, simply human "use and convenience" (Mitcham, 2009, 2020, pp. 365 ff.; Briggle, 2021, pp. 129 ff.). Some forms of growth are self-destructive, and often it is not easy to anticipate this in advance. But most importantly, it would seem necessary to face squarely the "use and convenience" end that has, since its emergence as a profession, become so deeply sedimented in the practice of English-speaking engineering.

3.4 Mythopoeic Oppenheimer: Apocalyptic *Bildung*

> In some sort of crude sense which no vulgarity, no humor, no overstatement can quite extinguish, the physicists have known sin; and this is a knowledge which they cannot lose.
> —J. Robert Oppenheimer, Massachusetts Institute of Technology, November 25, 1947

The classical theory and practice of *Bildung* focused on the person and assumed a stable world to which *Bildung* pursues a relationship. What does it become when the world becomes unstable? What does it become for those engineers who are intentional or unintentional contributors to world instability? There are at least two options: One is to deny the fact by means of some typical engineering risk-cost-benefit assessment that argues destruction is outweighed by benefit. Yes, engineering designs and creates nuclear weapons, but it also increases the human life span. Engineers defend their work as transformative rather than destabilizing, as making the world a better place in which more people can pursue *Bildung*. A second is to attempt to transform the instability producing activity itself, to change not the world by engineering, but engineering "by the world," as it were, by looking into the world that engineering is creating and responding appropriately, to create a kind of counter-engineering, the most common term for which is sustainable engineering.

This second approach is, I take it, the tact Buch and Christensen advocate. What they see in the engineered and engineering world is primarily an unintentional disruption of social justice. They thus propose to use Dewey to re-think engineering education in ways that will foreground its destabilizing or disruptive character and enable engineers to correct. While supporting their proposal, I want at the same time to table the possibility that engineering *disruptivity* (if I may coin a term) goes much deeper than they imagine and that counter-engineering may need to entertain more radical forms.

The inherent disruptivity was identified by engineer-scientist J. Robert Oppenheimer, who has taken on mythopoeic character. From the covers of *Time* (Nov. 8, 1948) and *Life* (Oct. 10, 1949) magazines, through popular books too numerous to catalogue, scholarly examinations (see especially Schweber, 2000 and Mason, 2006), a major biographical study (Bird & Sherwin, 2005), and now a block buster film (2023), Oppenheimer reveals to us things about ourselves that we both want and don't want to know. Once hailed as the man who release the energy of the sun and won World War II he was also dismissed as a "cry baby scientist" (by President Truman), targeted as a security risk, accused of being a communist, and deprived of his security clearance. During a famous 1947 Personnel Security Board inquiry into his argument against development of the hydrogen bomb the dual nature of engineering scientific research disruptivity was vividly on display. The disruptivity at issue concerns not only the physically disruptive impact in the form of massive destruction both real and potential, disruptions in military and political affairs national and international, and disruptions in the moral thinking of those intimately engaged with the relevant research and development—not just of nuclear weapons but of all types of the new category of socially destabilizing technoscientific projects. On April 13, in testimony about his initial attitude toward the atomic bomb project, he described how,

> When you see something that is technically sweet, you go ahead and do it and you argue about what to do about it only after you have had your technical success. That is the way it was with the atomic bomb. I do not think anybody opposed making it; there were some debates about what to do with it after it was made. (Polenberg, 2002, pp. 46–47)

A few days later, on April 16, under questioning that sought more detail about his attitude, there was the following revelatory exchange:

> Q. Why do you think you could have very well said that [the hydrogen bomb was a dreadful weapon]?
>
> A. Because I have always thought it was a dreadful weapon. Even from a technical point of view it was a sweet and lovely and beautiful job, I have still thought it was a dreadful weapon.
>
> …
>
> Q. You mean you had a moral revulsion against the production of such a dreadful weapon?
>
> A. This is too strong.
>
> Q. Beg pardon?
>
> A. That is too strong.
>
> Q. Which is too strong, the weapon or my expression?
>
> A. Your expression. I had a grave concern and anxiety.
>
> Q. You had moral qualms about it, is that accurate?
>
> A. Let us leave the word "moral" out of it.
>
> Q. You had qualms about it.
>
> A. How could one not have qualms about it? I know of no one who doesn't have qualms about it ….
>
> Q. But you still got on with the work, didn't you?
>
> A. Yes, because this was a work of exploration. It was not the preparation of a weapon…. It was an attempt at finding out what things could be done.
>
> …
>
> Q. Did you have moral scruples about [dropping the bomb on Hiroshima]?
>
> A. Terrible ones.
>
> Q. But you testified the other day … that the bombing of Hiroshima was very successful.
>
> A. Well, it was technically successful. (pp. 95–97)

Perhaps no one represents so openly struggles to come to terms with the moral political demands, both personal and political, inherent to the destructivity of contemporary science-based engineering and technology. Oppenheimer's engagements foreshadowed many ethical and political challenges since: in advanced medical life-saving and extending technologies, in genetic engineering (of plants, animals, and humans), in nano-bio-info-cognitive engineering, transhumanism, geoengineering, surveillance engineering, and in the artificial intelligence community. Seldom have subsequent engineers been as willing to be honest with themselves and with us about the fraught nature and stakes of our engineering prowess.

Much more common has been what Deep Mind founder Mustafa Suleyman, in his diagnosis of the coming wave of technology and power, terms "pessimism aversion" (Suleyman & Bhaskar, 2023). It's simply too difficult to think about and face potential catastrophe—except in escapist entertainment, where some superhero always saves the day. As Suleyman frames it, given the manifold of existential risks posed by the almost irresistible sweetness of generative AI and synthetic biology presents us with the simultaneous impossibility and necessity of containment.

Under conditions of potentially apocalyptic engineering, perhaps it would not be out of place to revisit the pre-Enlightenment meaning of *Bildung*: the teaching of persons as sinful creatures created in a divine image and, as such, called on to practice piety as a path to recovery of that imagine.

A final note about the meaning of "apocalypse": The Greek ἀποκάλυψις has two interconnected meanings. It can mean both destruction and revelation and even revelation through destruction. Not creative destruction but destruction that reveals the truth about the non-creativity of what once appeared to be creativity. Any consideration of the relevance of *Bildung* today, especially engineering *Bildung*, should begin by acknowledging the fragility and potential self-apocalyptic character of the engineered and engineering world.

3.5 Epilogue: Learning from Germany?

Now I want to make a frankly crazy thought experiment proposal for one way to enact or operationalize a response to the practice of *Bildung* under conditions of engineered destruction. The other country that most dramatically practiced the engineering of evil was Nazi Germany. My proposal invites consideration of drawing on the German response their own holocaust engineering. I do not mean to equate the atomic bombing mass destruction of Hiroshima and Nagasaki with the Shoah, but it is undeniable that both witnessed some measure of inherent engineering destructivity.

In her 2019 study *Learning from the Germans: Race and the Memory of Evil*, Susan Neiman, a Jewish American philosopher and director of the Einstein Forum in Potsdam, presents the German effort to accept responsibility for its World War II engineering atrocities—its *Vergangenheitsbewältigung* or "struggle of overcoming the past"—as providing a model for how Americans might deal with their historical atrocities against native Americans and Blacks.

As Deborah E. Lipstadt summarized the argument in a favorable *New York Times* review, Neiman

> contrasts Germany's response to the Holocaust with America's response to slavery and centuries of racial discrimination. Her concern is not "comparative evil" — which event is worse — but "comparative redemption," how each community has responded to and reframed the memory of its unsavory past. Neiman contends that postwar Germany, after initially stumbling badly, has done the hard work necessary to grapple with and come to terms with the legacy of the Holocaust in a way that could be a lesson to America in general, and the American South in particular. (Lipstadt, 2019, p. 14)

At the time of writing her book, Neiman thought Germany had to a significant degree accepted collective responsibility for its collective World War II crimes through what another scholar Dirk Moses (2021) subsequently termed "the German Catechism": "The Holocaust is unique … a civilizational rupture [such that] Germany has a special responsibility to Jews in Germany [and that] Israel's security is part of Germany's reason of state." (Neiman's thesis has been widely discussed and attacked and she has since reconsidered: See Neiman, 2023.) Nevertheless, I would like to put on the table the frankly crazy idea that engineers, or some subset of engineers, might adapt a version of that catechism: "The engineering apocalypse is unique, a civilizational rupture, one that engineers have an overriding responsibility to recognize and build into their professional self-consciousness and make part of their professional reason for being."

References

Bird, K., & Sherwin, M. J. (2005). *American prometheus: The triumph and tragedy of J. Robert Oppenheimer*. Random House.

Briggle, A. (2021). *Thinking through climate change: A philosophy of energy in the anthropocene*. Palgrave Macmillan.

Buch, A., & Christensen, S. H. (2025). Rethinking engineering professionalism through the concept of *Bildung*. In *Fill in*.

Christensen, S. H., Meganck, M., & Delahousse, B. (Eds.). (2007). *Philosophy in engineering*. Academica.

Dessauer, F. (1972 [1927]). Technology in its proper sphere. In C. Mitcham & R. Mackey (Eds.), *Philosophy and technology: Readings in the philosophical problems of technology* (pp. 317–334). Free Press.

Dewey, J. (1944 [1916]). *Democracy and education: An introduction to the philosophy of education*. The Free Press.

Florman, S. (1976). *The existential pleasures of engineering*. St Martin's Press.

Herdt, J. A. (2019). *Forming humanity: Redeeming the German Bildung tradition*. University of Chicago Press.

Lipstadt, Deborah E. (2019, August 27). Confronting evils of the past. *New York Times Sunday Book Review*, p. 14.

Mason, R. (2006). *Oppenheimer's choice: Reflections from moral philosophy*. State University of New York Press.

Mitcham, C. (2009). A historico-ethical perspective on engineering education: From use and convenience to policy engagement. *Engineering Studies, 1*(1), 35–53.

Mitcham, C. (2015). Ethics is not enough: From professionalism to the political philosophy of engineering. In S. S. Sethy (Ed.), *Contemporary ethical issues in engineering* (pp. 48–80). IGI Global.

Mitcham, C. (2020). *Steps toward a philosophy of engineering: Historico-philosophical and critical essays*. Rowman & Littlefield International.

Moses, A. D. (2021, May 23). The German Catechism. *Geschichte der Gegenwart*.

Neiman, S. (2019). *Learning from the Germans: Race and the memory of evil*. Farrar, Straus and Giroux.

Neiman, S. (2023, October 19). Historical reckoning gone haywire. *New York Review of Books*.

Polenberg, R. (Ed.). (2002). *In the matter of J. Robert Oppenheimer: The security clearance hearing*. Cornell University Press.

Schweber, S. S. (2000). *In the shadow of the bomb: Bethe, Oppenheimer, and the moral responsibility of the scientist*. Princeton University Press.

Suleyman, M., & Bhaskar, M. (2023). *The coming wave: Technology, power, and the twenty-first century's greatest dilemma*. Crown.

Carl Mitcham Visiting Professor of Philosophy of Technology at Renmin University of China and Emeritus Professor of Humanities, Arts, and Social Sciences at Colorado School of Mines. His work focuses on philosophy and policy issues related to engineering, technology, and science. His most recent books are a co-edited volume on *Philosophy of Engineering, East and West* (2018, with Li Bocong, Byron Newberry, and Zhang Baichun) and *Steps toward a Philosophy of Engineering* (2020).

Chapter 4
Bildung, Destruction, and Reconstruction: A Rejoinder to Carl Mitcham

Anders Buch and Steen Hyldgaard Christensen

Abstract In this chapter we discuss Carl Mitcham's reading of our suggestion to rethink engineering professionalism through the concept of *Bildung* and his proposal for apocalyptic engineering *Bildung*.

Keywords Carl Mitcham · *Bildung* · Apocalypse · Engineering professionalism · Optimism · Pessimism

Carl Mitcham's reflection (Mitcham, 2025, this volume chap. 3) on our vison to rethink engineering professionalism through the concept of *Bildung* (Buch & Christensen, 2025, this volume chap. 2) tempers expectations and critically reconsiders fundamental tropes of *Bildung*. Rightfully, he arrests optimist premises vested in *Bildung*-narratives that "…the world is well ordered, encompassing, and fundamentally good; the second that one's profession is engaged in practices to make things better." The calamities of the twentieth century, and the dire prospects of the twenty-first century, certainly gives Mitcham good reasons to challenge naïve presumptions about the world as an orderly place, and professional practices as bringing more order. Likewise, pessimist philosophers like Arthur Schopenhauer, Friederick Nietzsche, Oswald Spengler, and Emil Cioran have poignantly inflicted Western imaginaries of progress with wormwood and gall. Cosmic order and harmony are problematic ethico-ontological premises of *Bildung*-narratives and should not—in our view—guide visions.

This is yet another reason for rethinking engineering professionalism and *Bildung*. Our reconstruction proposal along John Dewey's philosophy stresses the

A. Buch (✉)
VIA University College, Valby, Denmark
e-mail: buch@via.dk

S. H. Christensen
Independent Researcher, Herning, Denmark

A. Buch, S. H. Christensen (eds.), *Bildung for Engineering Education and Practice*, Philosophy of Engineering and Technology 49,
https://doi.org/10.1007/978-3-031-86581-7_4

everchanging and precarious nature of human existence and the perils of surviving volatile, hostile environments. But it does not share the nihilism and fatalism of pessimist philosophers. It tries to balance between unwarranted optimism about progress and reason, and the crippling cynicism and despair of apocalyptic visions. This delicate act of balancing holds on to the *Sapere aude* motto of Enlightenment—Have the courage to use your own understanding—while humbly recognizing the fallibility and imperfectness of getting things right. It seeks to replace the arrogance and self-reliance of traditional Western Enlightenment thinking with aspirational approaches that are informed by values of care and respect for what is 'other', complex, and unknown (Buch & Juhl, 2018). This reflects the pragmatist plea to substitute the quest for certainty with unwavering social hope (Dewey, 1929/2008; Rorty, 1998) and to strive to ameliorate social problems, injustices, and suffering through thoughtful reflection, experimentation, and collective action. This is indeed a concept of *Bildung* for an ustable world.

Mitcham correctly sees that our proposal attempts not only to have engineers change the world to the better, but also to change engineering professionalism itself to become more responsive to and responsible for the everchanging world that we inhabit. However, he argues, the proposal might not be radical enough:

> Under conditions of potentially apocalyptic engineering, perhaps it would not be out of place to revisit the pre-Enlightenment meaning of *Bildung*: the teaching of persons as sinful creatures crated in a divine image and, as such, called on to practice piety as a path to recovery of that imagine. (Mitcham, 2025, p. 44)

Mitcham concludes that the engineering profession needs to acknowledge its complicity in the production of the atrocities of our past and present. Thus, he calls for a *Vergangenheitsbewältigung*—a struggle to overcome the past—in engineering, and the adoption of a 'catechism' that admits to and repent the misdoings of the past.

Surely, Mitcham's conclusion is meant to be provocative. It is also timely in correcting the 'pessimism aversion' (Suleyman & Bhaskar, 2023) in Western culture. However, we think that the radicalism of Mitcham's apocalyptic engineering *Bildung* suggestion needs critical attention before it totally discards the ideals of the Enlightenment project. Remorse and reflection on the past are needed for acting responsibly, but learning is fundamentally future-oriented (Dewey, 1909). Our reflections are always led by envisioning future states and concerns. Responsible action also needs *Zukunftbewältigung*—a struggle to face and overcome future challenges in intelligent ways. Thus, engineering professionalism must be rethought to meet the challenges of the future and become sustainable. Mitcham's pessimistic apocalyptical catechism should be mandatory reading in engineering curriculum, but it should not cripple the optimism of Enlightenment-thinking to build a better future. Our suggestion to rethink engineering professionalism is in fact quite radical.

References

Buch, A., & Christensen, S. H. (2025). Rethinking engineering professionalism through the concept of Bildung. In A. Buch, & S. H. Christensen (Eds.). *Bildung for engineering education and practice. A new agenda*. Springer.

Buch, A., & Juhl, J. (2018). *Sapere aude* revisited and revised. *Educational Philosophy and Theory, 50*(14), 1519–1520. https://doi.org/10.1080/00131857.2018.1458802

Dewey, J. (1909). *How we think*. D.C. Heath & Company.

Dewey, J. (1929/2008). In A. Boydston (Ed.), *The quest for certainty* (The later works) (Vol. 4). South Illinois University Press.

Mitcham, C. (2025). Bildung under constructions engineered destruction. In A. Buch & S. H. Christensen (Eds.) *Bildung for engineering education and practice. A new agenda*. Springer.

Rorty, R. (1998). Failed prophecies, glorious hopes. In R. Rorty (Ed.), *Philosophy and social hope* (pp. 201–209). Penguin Books.

Suleyman, M., & Bhaskar, M. (2023). *The coming wave. Technology, power, and the 21st century's greatest dilemma*. Crown.

Anders Buch Docent at VIA University College, Denmark and affiliated researcher at Jönköping University, Sweden. He holds an MA in philosophy from University of Copenhagen and a BEd and PhD in education from Roskilde University. His empirical research area is focused on professions, professionalism, and technological expert cultures, and his theoretical approach is primarily inspired by Science & Technology Studies, Practice Theory, and classical American Pragmatism. He has published articles and books on knowledge, learning, education, professionalism, and the professional development of engineers. Since 2016 he has servedas chief editor of Nordic Journal for Working Life Studies.

Steen Hyldgaard Christensen († 23rd June 2024). Independent researcher. He held an MA in Danish Language and Literature and the History of Ideas from Aarhus University, and PhD in research on engineering education from Aalborg University. Until 2014 senior lecturer at Aarhus University, Herning. From 2014 until May 2023, he was an adjunct associate professor at Aalborg University. He was initiator, coordinator, lead editor, and co-author of several edited volumes on engineering education, and epistemology. Five of them are published in the Springer Nature (POET) series. He has published articles on engineering epistemology, culture, and higher education. His research interests included the dynamics in higher education restructuring, and research initiatives to broaden engineering education.

References

Chapter 5
From the Individual to the Multi-agential Development of *Bildung* in Engineering Education through Challenge Based Learning

Diana Adela Martin, Andreas Spahn, and Gunter Bombaerts

Abstract In its day-to-day use, the German notion of *Bildung* can be translated as education. There is also a philosophical understanding of *Bildung*, as a concept that refers to the cultivation of human traits and capacities, as well as to the goal of this process. How can we understand *Bildung* as a goal of contemporary engineering education, and how does Challenge Based Learning serve as a pedagogical approach to accomplish it? To answer these questions, our contribution starts with an exploration of five interpretations of *Bildung* rooted in the history of philosophy, beginning with a focus on (1) self-determination, (2) self-understanding, (3) inner motivation, and (4) individual autonomy as key dimensions in the development of *Bildung* in the eighteenth and nineteenth century, followed in the twentieth century by a focus on (5) intersubjectivity. We proceed with a reflection on recent trends in engineering education to help us reposition *Bildung* both as an individual and multi-agential notion. We suggest that Challenge Based Learning can cultivate both the individual idealisation presupposed by the notion of *Bildung* as initially conceived and the relational focus presupposed by multi-agent interactions and collaborative endeavours witnessed currently in engineering education, which is closer to the development of *Bildung* as intersubjectivity. We offer three educational examples where Challenge Based Learning exposes students to real socio-technical challenges put forward by stakeholders from the local ecosystem of the university. We discuss the implications of bringing together the individual and multi-agential understandings of *Bildung* in engineering education in connection to these examples. The conclu-

D. A. Martin (✉)
University College London, London, United Kingdom
e-mail: diana.martin@ucl.ac.uk

A. Spahn · G. Bombaerts
Department of Industrial Engineering and Innovation Sciences, TU Eindhoven, Eindhoven, The Netherlands
e-mail: a.spahn@tue.nl; g.bombaerts@tue.nl

A. Buch, S. H. Christensen (eds.), *Bildung for Engineering Education and Practice*, Philosophy of Engineering and Technology 49,
https://doi.org/10.1007/978-3-031-86581-7_5

sion offers a reflection on the prospects and possibilities for *Bildung* in engineering education.

Keywords Challenge based learning · Relationality · Self-determination · Inner motivation · Self-understanding · Autonomy · Intersubjectivity

5.1 Historical Interpretations of *Bildung*

Johann Gottfried von Herder is credited with putting forward the first coherent pedagogic vision of *Bildung* (p. 37). Its focus was on developing the capacities of the individual in the spirit of humanistic values. The eighteenth century witnessed growing discussions and articulations of the notion of *Bildung*, which saw scholars such as Immanuel Kant, Friedrich von Schiller, and Johann Friedrich Herbart all contributing towards its articulation. The institutionalization of *Bildung* as an educational programme came via Wilhelm von Humboldt, whose aim was to reform the university system (Backhaus, 2015).

In its day-to-day current use, the German notion of *Bildung* is typically translated as education. It is a process of reverse translation, as the notion of *Bildung* made its way into the German thought of the eighteenth century as a translation of what Anthony Ashley Cooper, the third Earl of Shaftesbury, called in his moral writings the "inward form" (Horlacher, 2004). At its origins, *Bildung* had a political dimension linked to the development of universities as institutional locations for learning that are separated from the Church or ideological interferences (Habermas, 1991; Östling, 2018). In this way, *Bildung* has been closely linked to the emergence of citizenship and national consciousness in the German civil society (Elias, 2010). There is also a philosophical dimension of *Bildung* as a concept that refers to the cultivation of human traits and capacities, as well as to the goal of this process (Nordenbo, 2002). Nevertheless, the term *Bildung* as developed in the German tradition, has no direct counterpart in English, which comes with consequences for international educational discourses (Varkøy, 2014). According to Varkøy (2014), "even today's European discourse of knowledge and education lacks a functioning concept of *Bildung* as something that is qualitatively different from education". To compound the difficulty of translating and understanding the notion of *Bildung*, Biesta acknowledges (2002, p. 344) that "there is no such thing" as *Bildung*.

Given the vague and indeterminate nature of the notion of *Bildung* (Bohlin, 2008), we start by disentangling the various ways in which it has been articulated. Historically, *Bildung* has been transposed into four key theoretical tenets: (1) self-realisation/determination, (2) self-understanding, (3) inner motivation, and (4) individual autonomy. These tenets have been identified inductively through a critical reading of key authors associated with the development of the notion of *Bildung*.

5.1.1 **Bildung** *as Self-determination*

Self-determination is a process of intentional and continuous personal development. It involves nurturing one's intellectual, emotional, and ethical aspects through various practices, experiences, and reflections (Pinkard, 2010). The ideal of individual autonomy and the self-determination of the learner have been central to the German development of the notion of *Bildung* in Germany between 1770 and 1815 (Alves, 2019). Notable is Herbart's (1806/1908, p. 62) understanding of self-determination (*Selbstbestimmung*) as the ability of the moral individual to "command himself." The human being is seen by Herbart (1802/2022) as a learning being:

> The subject of education is without any doubt not man in the most general sense, but man as a changeable being, as a being that can make transitions from one state into another, while yet capable of maintaining something of himself in the new situation.

Bildung as self-determination is also manifest in the work of von Humboldt (1793/2000), who emphasizes achieving "as much substance as possible for the concept of humanity in our person" as "the ultimate task of our existence." He believed that education should allow individuals to freely pursue knowledge and intellectual growth, which would lead to self-determined actions in their personal and professional lives.

Self-determination is also seen to align closely with the idea of self-discovery and self-realization. In "Letters for the Advancement of Humanity," Herder (2002, pp. 374–379) advocated for education that nurtures unique talents and capacities, allowing individuals to pursue their own path in life. According to Berlin (1976), Herder believed that self-determination could be achieved by encouraging learners to explore their interests and engage with diverse cultural traditions.

5.1.2 **Bildung** *as Inner Motivation*

Bildung has also been conceived as inner motivation. Inner motivation refers to the inherent drive and desire within an individual to engage in learning, exploration, and self-improvement. It is driven by curiosity, autonomy, and the desire for personal growth and fulfillment.

According to Herbart (1806/1908, 219), inner motivation is linked with the role of education in fostering and broadening the interests of the learner, together with the belief that learning is, in turn, fostered through educational experiences that are of interest to the learner:

> I am, however, convinced that the proper hardening principle for man [...] will not be found until we learn how to arrange a mode of life for the young, whereby they can pursue, according to their own and indeed, their right mind, what in their own eyes is a serious activity.

As such, for Herbart, inner motivation is driven by the learner's inherent curiosity and interests, stemming from the individual desire to make sense of new information based on prior knowledge and experiences (Schiefele, 1991). The role of education went beyond transmitting information and included the focus on developing a learning environment that would foster curiosity and the desire to learn. Developing inner motivation was for Herbart a moral task (Valdemarin, 2022). He considered that inner motivation can be guided by the learner's moral disposition, in turn contributing toward virtuous actions.

Humboldt shared the belief that education should aim to stimulate and nurture individuals' inner motivation. In *The Limits of State Action* (von Humboldt, 1793/2000), he highlighted the link between learning environments that foster intellectual freedom and the inner motivation of learners, who become self-directed in their pursuit of knowledge and personal growth. This aligns with the idea that inner motivation is fundamental to *Bildung*, as learners become active participants in their own education.

5.1.3 **Bildung** *as Self-understanding*

The interpretation of *Bildung* as self-understanding in German thought is considered to have been influenced by the translation of Shaftesbury's moral works. Shaftesbury proclaimed that "the hardest thing in the world is to be a good thinker, without being a strong self-examiner" (Shaftesbury, 1710/1981, p. 58). For Shaftesbury, self-improvement requires self-reflection, an idea that made its way via Herder to Humboldt (Kauffmann, 2012, p. 34). Both Herder and Humboldt considered self-understanding an essential aspect of *Bildung*, which relies on the development of an individual's sense of identity, cultural awareness, and moral sensibility (Eck, 2019).

For von Humboldt (1793/2000), the reflective process of self-understanding is conceived to facilitate self-determination:

> In pure, ultimate terms, thought is never more than an attempt of the mind to be comprehensible to itself [...] The procedure of our mind, particularly in its more mysterious effects, can be fathomed only by deep reflection and unceasing observation.

Bildung is thus seen as a way to become freer due to enhanced self-reflection. This is also a core idea of the literary genre of the *Bildungsroman*, whose core proponent is Johann Wolfgang von Goethe. In *Wilhelm Meister's Apprenticeship,* the process of self-discovery is facilitated by reflecting on one's mistakes in order to overcome them. Herbart accentuates the role of inner conflict as a moral basis of education. For him, self-understanding sees the learner expanding their horizon and distancing from their initial inclinations (English, 2014).

5.1.4 Bildung *as Individual Autonomy*

According to Humboldt (1850/1993), the *Bildung* ideal of cultivating the inner self presupposes the autonomy of the individual:

> whatever does not spring from man's free choice, or is only the result of instruction and guidance, does not enter into his very being, but still remains alien to his true nature.

Individual autonomy refers to a person's ability to think independently, exercise personal agency, make choices that align with their own ideals and principles, and take responsibility for their actions and decisions. In Humboldt's view, as individuals become educated and develop their intellect, they gain the tools to exercise autonomy in their decision-making processes and actions. This self-directedness is seen as essential for personal fulfilment, social participation, and developing into a responsible member of society.

According to Humboldt (1850/1993), education should be oriented toward fostering individuality and autonomy. He envisioned a liberal education that empowers learners by focusing on developing individuals' intellectual faculties and enabling them to think critically and independently. In his educational reforms, Humboldt (1850/1993), proposed that universities should be autonomous institutions, free from state interference, and committed to the pursuit of knowledge for its own sake. Humboldt's vision of a university emphasized the importance of free inquiry and academic freedom, enabling learners and teachers alike to engage in self-directed learning.

5.1.5 Bildung*'s Turn Towards Intersubjectivity*

What these four initial interpretations of *Bildung* have in common is a focus on the learner *qua* individual within the education system. It was after the second world war that philosophical discussions of *Bildung* put forward the embeddedness of this notion in an intersubjective context. This idea developed in two very different schools of thought: in Gadamer's traditionalistic revitalisation of the concept of *Bildung,* on the one hand, and in the re-interpretation of *Bildung* as critical thinking within critical theory in Habermas and Apel, on the other hand. Even though both interpretations of *Bildung* are obviously very different, it is striking to see that both traditions develop a relational, intersubjective conceptualisation of *Bildung* (Spahn, 2009).

Gadamer rejects the starting point of the transcendental subject in Kant, that has become very influential in German philosophy. He argues that the humanities are better advised to return to concepts in their tradition that have been neglected, such as *Bildung* and the *sensus communis*. In *Wahrheit und Methode* he develops the argument that the humanities have been set on the wrong foot ever since the Cartesian dualism that takes the subject and the object as the two central categories of ontology but ignores the relation between multiple subjects (Gadamer, 1989). This echoes the criticism of earlier thinkers, such as Dilthey (1910/2021), that Kant did not take the humanities seriously, so that what is needed is to write an additional *Kritik der verstehenden Vernunft* (Critique of Interpretative Reason) to counter the thread the positivism poses to the self-understanding of the humanities. Gadamer himself emphasises the importance of two relational concepts: the *sensus communis* and the authority of *tradition* in which we stand (Gadamer, 1989). Whereas Kant tried to guarantee a philosophical foundation by appeal to a common timeless ideal of human rationality, Gadamer rejects this a priori approach, since it ignores the historicity and relationality of truth. He refers to thinkers such as Vico, Hegel and Herder, who in different fashions emphasised the societal embeddedness of knowledge (Gadamer, 1972). Truth in the humanities, and thus in *Bildung*, is for Gadamer not linked to categories of abstract methodological reasoning, but linked to an artful dialogue with the tradition in which we are situated and an acknowledgment of the situatedness in a community (Gadamer, 1953).

Critical theory scholars reject this seemingly uncritical embracing of the tradition in Gadamer since it seems to be incompatible with the emancipatory interest of critical theory. However, also within critical theory, we find an emphasis on the importance of relational categories, often termed in the terminology of intersubjectivity (Apel, 1998, Habermas, 1981). Apel and Habermas even develop the thesis that the history of Western Philosophy as such follows the three paradigms of object-, subject- and intersubjectivity (Apel, 2011, Habermas, 2019). Apel argues that the Object, Subject and Intersubjective philosophy can be regarded as the three main historical paradigms of Western Philosophy (Apel, 1998). He argues that the understanding of being is the central task for Ancient and Medieval philosophy, making ontology and *metaphysics* the first philosophy.

Furthermore, humans are not seen as radically opposed to the world, but as embedded in either the cosmos or in God's creation. In modernity, philosophy turns more and more towards the individual subject as the central category and away from ontology. Apel discusses the examples of the cartesian cogito, Kant's Copernican revolution and his intentionalism in ethics, and Husserl's emphasis on the transcendental ego. Together with this turn *epistemology* becomes the most important philosophical discipline. Apel and Habermas argue finally that contemporary philosophy needs to take intersubjectivity as its starting point. The linguistic turn in analytic philosophy, the focus on the role of language both in continental and analytic philosophy and the rise of pragmatism are seen as examples of a growing engagement of philosophy with relational and intersubjective categories. Apel himself therefore aims to transform the Kantian Philosophie towards an intersubjective version of key enlightenment ideals, putting the embedding of individuals in communicative relations central. Similarly, Habermas stresses in his work the importance of *communicative* rationality, and the importance of the participant perspective also for the theoretical work of social theorists themselves (Habermas, 1987). This turn toward relational or intersubjective philosophy has consequences for a contemporary interpretation of the notion of *Bildung*, which we will explore in the next sections.

5.2 Contemporary Engineering Education: In Search for *Bildung*

Before attempting to frame *Bildung* as a contemporary ideal of engineering education, it is important to first explore recent trends in the development of engineering education and their societal anchoring.

In today's rapidly changing world, engineers are faced with complex problems such as climate change, sustainable development, or urbanization, which cannot be effectively addressed single-handedly or through a narrow disciplinary approach. For instance, the Grand Challenges for Engineering identified by the National Academy of Engineering (2017) call for interdisciplinary solutions to global issues, including access to clean water, renewable energy, and improved healthcare. To equip students with the necessary skills to tackle these challenges, engineering programmes have been integrating interdisciplinary coursework and collaborative projects that involve diverse stakeholders from different fields. Engineering education has thus become increasingly collaborative and oriented toward external agents (Diaz Lantada, 2020). Such collaborative approaches extend beyond the classroom, through the forging of partnerships with external organizations. This is exemplified by universities collaborating with local industry, start-ups, communities, or NGOs to provide students with hands-on experiences and exposure to real engineering projects. As such, we are witnessing in the past decade the emergence and sedimentation of outwardly facing forms of education, which include both curricular formats such as Challenge Based Learning, Project Based Learning, Design Based Learning, Place Based Learning, Service Learning and Humanitarian Engineering, as well as extra-curricular options, such as community design competitions or industry placements (Martin, Bombaerts, et al., 2023).

This shift in engineering education can be attributed to the recognition of the value of experiential learning and its potential to enhance students' skill sets and their adaptability to real-world challenges. For instance, the CDIO (Conceive, Design, Implement, Operate) Initiative promotes project-based learning where students work in interdisciplinary teams to address authentic engineering problems, often collaborating with industry partners to address practical problems (Crawley et al., 2007). In service learning and humanitarian engineering we see collaborations with communities with the aim of identifying and implementing solutions that address pressing social and environmental issues (Mazzurco & Daniel, 2020). Humanitarian engineering and service learning provide avenues for engineering students to integrate technical, social, cultural, and environmental considerations and develop community-oriented solutions (Mitcham & Munoz, 2010). The connection with the local site is especially important in place-based learning, where students get to engage with the physical and cultural contexts of specific locations to gain a deeper understanding of engineering principles and real-world applications (Yemini et al., 2023). This approach emphasizes the integration of local knowledge, community needs, and environmental factors into the engineering curriculum, fostering a sense of relevance and connection between engineering concepts and their practical implications (McGowan & Bell, 2020). Additionally, we witness the emergence of engineering programs that emphasize design thinking methodologies, encouraging students to engage with diverse stakeholders and develop solutions that cater to users' needs (Bertoni, 2020). This approach focuses on fostering an understanding of the needs and contexts for which engineering students are designing solutions (Brown & Martin, 2015; Platttner et al., 2010).

The growth of pedagogical approaches that make use of experiential learning, multidisciplinary teamwork, and stakeholder collaboration—implemented either at the course or program level—show how contemporary engineering education actively involves external agents to create a more engaging and relevant learning experience for students, with a view towards preparing them to be adaptive and prepared for practice (Martin et al., 2021b).

At the same time, we can still identify in engineering education the ideal of *Bildung* as articulated in the eighteenth and nineteenth-century philosophy via the current discourse about the need to develop holistic and well-rounded graduates capable of exercising their societal responsibility and applying their technical skills to societal issues (Mitcham, 1994). At the classroom level, this translates into curricular content that includes ethics, virtues, empathy, social justice, decolonization, and global and cultural awareness (Cordeiro Cruz, 2021; Leydens & Lucena, 2017; Martin & Conlon, 2023; Martin, Gwynne-Evans, et al., 2023; Niles et al., 2020). As engineering students advance in their education, they develop professional expertise and autonomy (Reyes Viviescas et al., 2019) and "undergo changes in their identity and self-conception of what it means to be an engineer" (Pierrakos et al., 2009).

5.3 A Relational Repositioning of *Bildung* for Engineering Education

Given the growing collaborative and experiential nature of engineering education (Martin, Bombaerts, et al., 2023; Masschelein & Simons, 2002), we suggest the need to complement the individualistic understanding of *Bildung* with a multi-agential understanding that accounts for the relations of learning between individuals and/or collectives, as members of the same but also of different systems. In what follows, we will explore what a relational repositioning of *Bildung* implies.

According to English (2014, p. 14), *Bildsamkeit* refers to the process of change through *Bildung,* which sees individuals as "changeable beings." Here seems to lie a conundrum when thinking of *Bildung* as an individual process of change: the relational process of joint learning is lost, and so is its societal outcome of localized growth at the regional or national level.

To this individual-oriented lens, we propose understanding *Bildung* as a multi-agential process of change (Table 5.1). As such, we build forth on the turn towards intersubjective relations in both traditionalistic conceptualisations of *Bildung* and in

Table 5.1 A Relational Repositioning of *Bildung*

	Individual	Multi-agential
Subject of learning	The student	Academic and non-academic individuals and groups
Learning site	The educational system	The ecosystem in which individuals and collectives are situated
Positioning of the learner	Fixed within a close-bounded system (i.e. classroom or university)	Interactive within the university ecosystem
Learner agency	Internal determinants and independent of others	Manifest in group settings and oriented at the socio-economic, cultural, and political context

critical theory. This relational starting point is especially important when considering the prospects for *Bildung* in engineering education.

The institutionalization of *Bildung* in pedagogical forms, such as the case of engineering programmes offered by technological universities, needs to acknowledge the interactions purporting to learning that occur between different types of individual agents (including students, teachers, administrators) or collective agents (universities, schools, firms, community groups, NGOs a.o.), towards mutual self-realisation, motivation, understanding and autonomy. If we are to think of engineering education as contributing to the *Bildung* not only of individuals, but of the ecosystem in which technological universities are situated, we need a multi-agential and relational discourse.

The multi-agential development of *Bildung* is a moral endeavor; one that invites us to rethink the ontological and epistemic foundations of ethics. Taylor (2017) calls for "a view of ethics-in-relation," which according to her "does not sit well with the traditional assumption that *Bildung* happens within the *Bildungssystem*." Quoting Bennett (2010), Taylor appears to put forward a relational view of responsibility, as residing "in one's response to the assemblages in which one finds oneself participating".

5.4 From the Individual to the Multi-agential Development of *Bildung* via Challenge Based Learning

In what follows, we will present three case studies set at TU Eindhoven that present the multi-agential integration of *Bildung* in engineering education via Challenge Based Learning (CBL). Challenge Based Learning is a pedagogical approach that brings to the engineering classroom real-life problems based on the activity of real-life stakeholders (Kohn Rådberg et al., 2020). These are presented to students as socio-technical challenges embedded in a learning context (Martin & Bombaerts, 2024). According to a recent review, CBL combines scientific content areas with societal considerations, under the format of challenges of global importance that students need to address (Gallagher & Savage, 2020).

In CBL, external stakeholders play a key role and serve as differentiators from other pedagogical approaches such as project based learning (Hadgraft & Kolmos, 2020). An external stakeholder can be R&D centers and laboratories, start-ups, SMEs and large industry companies, NGOs, citizen associations, local communities, research groups, consumer and user groups, local and national governments, think-tanks, national or international policy bodies, environmental associations, groups and networks, humanitarian bodies, underprivileged or at-risk communities, and communities affected by natural or man-made disasters (Martin, Bombaerts, et al., 2023). These stakeholders ask students to help solve a challenge they face or work on themselves. The challenges are broad problem areas that students need to narrow down to a specific problem that they will then address. In CBL, students are not only problem-solvers, they are also problem-definers (Martin & Bombaerts, 2024).

5.4.1 Challenge Based Learning at TU Eindhoven

TU Eindhoven strives toward a curriculum-wide adoption and implementation of CBL (Rijk, 2019). At present, there are 40 CBL courses offered both at Bachelor's and Masters' levels to more than 4000 students. We describe three courses that implement CBL where the authors have been involved as designers, teachers and/or researchers. Based on our experience, we will then discuss how the individual and multi-agential dimensions of *Bildung* can be understood and pursued in this educational format.

Case Study 1: Decision Under Risk and Uncertainty

The first case is a second-year course sequence on decisions under risk and uncertainty comprised of 3 courses of 5 ECTS unfolding over a period of 30 weeks. The first course in the sequence was dedicated to ethical issues related to participatory techniques in technological development, social justice, and risk awareness. The second course in the sequence brought technical knowledge about risk assessment techniques. The third course in the sequence was a project course that required students to develop a risk analysis that incorporates ethical and technical aspects. This is an interdisciplinary course, with 77 students, 14 teachers and 7 stakeholders, with an additional 4 guest speakers. While the majority of students followed the Chemistry and Chemical Engineering degree, other disciplines represented were Mathematics, Electronics, Computer Science, Industrial Engineering, Physics, Philosophy and Technology. The stakeholders comprised two start-ups originating in students teams based at TU Eindhoven, the organisation Engineers Without Borders Netherlands, a regional waterboard, a research group working on cryptography, a research group working in collaboration with a local hospital on plasma medical solutions and a researcher working with a company on hydrogen technology. Thus, students could choose one of 7 real-life projects, ranging from the transformation of plastic waste into bricks in Africa, developing preventive or reactive technologies against battery fires, increasing the biogas production from wastewater treatment plants in`s-Hertogenbosch and extending the facility with an ammonia stripper, evaluating the safety of hydrogen steam reforming, introducing plasma medicine technologies on the market, analyzing solutions for Privacy-Preserving Authentication in the comment section of media sites and the development of personalised cognitive development training programs. The course gave students the opportunity to reflect on their professional development as responsible engineers and how to ensure just and sustainable technological solutions to societal problems.

Case Study 2: E3 Challenge2 Ethics and Data Analytics
The second case is a first-year 8 weeks course of 10 ECTS on data analytics and the ethics of technology. This is an interdisciplinary course, with 45 students and 1–3 stakeholders. In the academic year 2022–23, the stakeholder is a research group based at TU/e which examines Brain on Chip (BoC) technologies. Students receive a complex data set from the research conducted on BoC and have to analyze it. The BoC research group ask students to propose a sociotechnical solution in the form of data analysis that also addresses ethical issues. Previously, students took an introductory course on data analytics. In their data analysis, students include ethical considerations purporting to the privacy of stem cell usage for medical applications, informed consent, moral enhancement, brain organoids, and animal testing replacement.

Case Study 3: Physics of Social Systems
The third case is a second and third-year 24 weeks course of 15 ECTS on the physics of social systems. This is an interdisciplinary course that brings together physics, psychology and the ethics of technology. There are 45 students and 1–3 stakeholders. In 2022–23,the stakeholder was the Dutch railway operator ProRail, which is responsible for the maintenance and extension of the national railway network infrastructure, the allocation of rail capacity, and controlling rail traffic. As part of their activity, ProRail is conducting research on pedestrian movement in train stations. Students receive a complex data set from ProRail and have to model pedestrian movements, set up a nudge in a Dutch railway station, and measure the outcomes of their nudge. There are ethical aspects encountered at all stages of the development of a solution. First, there are ethical aspects linked to informed consent, as students have to write their own ethics review board approval before conducting their study. Second, students need to interpret from the data how pedestrians show ethical values to each other, such as responsibility, care, safety and comfort, and include these considerations in the model. Third, students have to reflect on how they can use these insights to propose more ethical ways for crowd nudging.

5.4.2 *The Individual Dimension of* Bildung *in CBL*

Research conducted at TU Eindhoven found that CBL fostered the self-determination, inner motivation and autonomy of the learners (Bombaerts, 2020, 2021; Bombaerts et al., 2018, 2021, 2022; Bombaerts & Nickel, 2017; Bombaerts & Vaessen, 2022).

In psychology, self-determination refers to an individual's ability to autonomously make choices and decisions that align with their own values, interests, and goals, and encompasses various dimensions. Self-determination theory posits that fostering basic needs such as autonomy, competence, and relatedness supports intrinsic motivation and psychological well-being (Ryan & Deci, 2000). CBL can contribute to the enhanced self-determination of learners in small-scale course settings (Bombaerts, 2020). It also contributes to the inner-motivation and autonomy of students by providing learners with opportunities to choose projects aligned with their interests and to define by themselves the problem they work (Martin & Bombaerts, 2024). This is seen to promote among students a sense of ownership and autonomy over their learning journey. The autonomy they receive enhances self-motivation, as students are intrinsically driven to explore the topics they are passionate about (Bombaerts, 2020). The real-life societal relevance of the projects and the interaction with stakeholders also motivate learners to apply their knowledge and skills to make a positive difference in their community (Bombaerts et al., 2021). Seeing the practical value of their learning when addressing complex societal problems leads to higher student engagement and greater intrinsic motivation (Martin & Bombaerts, 2024). To increase motivation, CBL courses also include moments of brainstorming to inspire innovative ideas that students believe are relevant to pursue. CBL contributes to enhanced self-understanding by instilling regular reflection practices (Bombaerts et al., 2022).

Summing up, the main strategies by which CBL fosters the individual dimension of *Bildung* is by providing students with opportunities to choose projects and define them, allowing them to set their own learning goals and monitor their own progress, encouraging self-directed learning, spotlighting the real-life value of their course work and assignments, and creating opportunities for reflection.

5.4.3 *The Multi-Agential Dimension of* Bildung *in CBL*

Considering the categories in Table 3.1, we can now observe how these play out in CBL, contributing to a relational repositioning of *Bildung*.

5.4.3.1 The Subject of Learning in CBL

In the initial development of *Bildung*, the student is considered the subject of learning and personal growth. This view is shared by traditional forms of education. In CBL, students get to interact closely with a range of external stakeholders, and this interaction can represent a meaningful learning experience not only for the students but for the stakeholders as well. This collaborative approach fosters the exchange of ideas between students and stakeholders. While students learn how to tackle a sociotechnical problem and develop practical skills and disciplinary knowledge, stakeholders are exposed to fresh perspectives and innovative solutions, sometimes outside their original area of concern.

For stakeholders to be also learners in CBL, the educational process benefits by having a sufficiently high number of student groups working on their challenge, to enhance the chance of being exposed to out of the box ideas. As students are exposed to societal content and ethics, this becomes a concern that translates into their projects and is passed on to the stakeholders themselves. As such, the interaction with students in CBL can bring in ethical elements that are useful for the external stakeholder and enhance their awareness of their societal responsibility. The more open the challenge, the more unplanned the co-learning process. Stakeholders gain access to a diverse talent pool and can leverage students' creativity to learn new perspectives that can constitute innovative solutions in their line of work. In some cases, where students consider that the stakeholders are not interested in ethical aspects but only in a technical fix to a problem, they can be quite critical in pointing out the organisational weakness of the stakeholders. The teacher has to manage the process of learning on both sides, to be beneficial for both students and stakeholders, without making the stakeholders too uncomfortable with the criticism received.

5.4.3.2 The Learning Site in CBL

CBL supports a diverse range of locations and learning sites, each offering unique characteristics and prospects, that make use of the entirety of the university ecosystem. Learning sites are structured to encourage collaboration and interaction, for students to pool their diverse perspectives, skills, and knowledge to develop their sociotechnical projects.

First, there is the physical space offered by the university campus. Here, students have contact moments in rooms whose furniture positioning and design facilitate group work and discussions with peers, teachers and external stakeholders. Students can also work in labs or maker spaces. Notable is the Innovation Space (TU Eindhoven, n.d.), which has special equipment and facilities (such as equipment for 3D printing) as well as workshops and workspaces to allow the development of prototypes in an interdisciplinary group setting.

Second, in CBL students can also make use of the location of the companies to develop the technological parts of their challenge, receive tours explaining the technical processes involved in the activity of the stakeholder or receive targeted training (i.e. on safety workplace techniques or the use of specific equipment).

Thirdly, students can also make use of online environments that stimulate discussion and collaboration outside university hours.

5.4.3.3 The Positioning of the Learner in CBL

In CBL, learning is interactive. Students are learning with and alongside their peers, their teachers, external stakeholders or the communities they represent.

First, CBL encourages student-student interactions. As the problem areas addressed via CBL are interdisciplinary in nature and with a strong societal component, students work in teams often comprising different disciplinary backgrounds and skills. The interactive elements of learning are central when articulating the problem to be solved and then in addressing it. Students get to negotiate between themselves their contribution to the project and balance the deployment of skills they are already strong at with skills they don't possess and wish to develop.

Second, learning occurs also in the interactions between students and teachers. In their interactions, teachers bring to the forefront the societal dimension of the project and focus on the ethical development of students. This is an intense learning experience often achieved through coaching or mentorship sessions. In such sessions, the teacher provides in-depth developmental feedback and encourages students to engage in self and group reflection (Bombaerts et al., 2022). Through these learning activities, students get to develop their ability to reflect on their own personal and professional values and how to incorporate them in their day-to-day practice and their projects. Furthermore, students also come to regard their interpersonal interactions as a source of morality in itself, when addressing aspects such as their own responsibility towards their team-mates, the impact of missing deadlines set by the group, creating an inclusive team, encouraging equal contribution or knowing when to speak up and highlight problematic issues.

Thirdly, learning occurs in student-stakeholder interactions. In CBL, students are encouraged to conduct research to better understand the technical and ethical issues related to the problem area. This involves consulting academic articles, news outlets or video lectures. Research is essential in coming up with a relevant problem and framing its ethical dimension and societal implications. At different moments in the course, students meet with stakeholders an get their feedback. There are typically 2–3 meetings for a course lasting 10 weeks. In an initial meeting, stakeholders typically introduce the challenge and answer any questions students might have. An intermediary meeting may be scheduled for the middle of the course, with students presenting their initial proposal, for which they receive feedback from the stakeholder. The final meeting usually takes place in the last week of the course, when students deliver their proposed solution and clearly communicate the ethical principles guiding their decision-making. In the meetings, there is ample opportunity for dialogue between the students and stakeholders. By engaging with ethical dilemmas throughout the span of the project and communicating to the stakeholder the need to consider ethical aspects in their work, students can develop their ethical decision-making skills, learn to apply ethical principles in complex situations and argue in favor of an ethical solution in the concrete interaction with external stakeholders and mindful of the characteristics of their surrounding context.

5.4.3.4 Learner Agency in CBL

In CBL, the learner agency is manifest in group settings and oriented at enhancing the socio-economic, cultural, and political context through the projects developed or the awareness created about the potential impact of technological projects and innovations.

While the CBL deliverables are embedded in an educational context, students get to engage with problems external to the university, with the aim of developing ethical solutions and preparing students to be change agents. The projects students work on in CBL are often connected to real-world problems. When students engage with these problem areas, they understand that their solutions can have tangible impacts on their communities, the environment and overall on society. This awareness of real-world consequences encourages them to take their work seriously and situate the outcomes in a specific context.

Learners are aware of their situatedness within a specific socio-economic, cultural, political context, and considerate of the distinct features and characteristics of the external agents (users, consumers, communities a.s.o.) that can be potentially impacted by their work and engineering projects.

Most projects in challenge-based learning are centered around societal or environmental issues. Students are encouraged to identify those potentially affected by a project, develop empathy and take social responsibility seriously. This sense of empathy and social consciousness nurtures the commitment to make a positive impact on the world. For example, one of the projects in the Decisions Under Risk and Uncertainty course has as a problem area the issue of plastic waste in Africa and its conversion into cost-effective building materials, such as plastic bricks. Students had to identify three categories of stakeholders, who may be the risk exposed, beneficiaries and decision-makers, and frame a sociotechnical solution to the problem of plastic waste and a method for achieving it that would turn the groups identified as risk exposed also into beneficiaries and/or decision-makers. Students can feel empowered when their coursework strives to alleviate a real-life problematic aspect. Furthermore, they also learn about participatory and co-creative methods that can also empower the communities served by engineering projects, and makes them partner in the process. CBL facilitates educational interventions and learning experiences where students and external stakeholders get to work together in a way that their solution makes a societal difference.

5.5 Conclusion

The article contributes to recent debates about the role and relevance of *Bildung* in contemporary education (Heidt, 2015), in the context of engineering education (Christensen et al., 2006; Christensen et al., 2007; Gaidi, 2009). Following other scholars who noted the historical transformation of the notion of *Bildung* (Hamann, 2011), we traced its development from individualism to intersubjectivity, and explored its implications for engineering education. We described how contemporary engineering incorporates teaching approaches and ideals that resonate with both the individual and multi-agential dimensions of *Bildung*. Engineering education experiences the growth of collaborative and interdisciplinary methods that are reflective of the complex nature of societal challenges engineers solve. At the same time, through a focus on ethics and societally-oriented content, engineering education strives to foster the development of the professional identity of engineering students as responsible professionals and socially engaged agents (Martin et al, 2021a). To account for the collaborative nature of contemporary engineering education, we put forward a lens to help us situate the notion of *Bildung* relationally. Our contribution argued that Challenge Based Learning can accommodate both the individual and multi-agential dimensions of *Bildung,* and we offer examples set at TU Eindhoven. In the context of the CBL education offered at TU Eindhoven, we discussed how CBL fosters the individual dimension of *Bildung* by enhancing the self-determination, inner motivation, self-understanding, and autonomy of its students. We also looked at how the multi-agential dimension of *Bildung* is rendered via CBL at TU Eindhoven, looking at how the subject of learning is understood, the characteristics of the learning site, the positioning of the learner and the learner agency.

Our chapter has two main contributions. First, we strived to enrich the theoretical understanding of *Bildung* by offering a relational framing of *Bildung* that aligns with its key historical dimensions and its development. Second, we aimed to offer inspiration to educators who seek to foster *Bildung* in their teaching, by offering practical examples and a discussion of how *Bildung* can be cultivated via Challenge Based Learning, considering our own positionality as teachers and educational researchers.

We want to encourage educators to experiment with CBL in their teaching, and we put forward the following recommendations:

1. To start on the CBL path, you can consider reaching out to representatives of local communities, major employers in the area, colleagues who are overseeing student entrepreneurial projects or initiatives at the department or university level or your national chapter of Engineers Without Borders. These may be potential partners to offer projects linked with their activity. It is important to discuss expectations and set realistic time commitments, as well as to identify any potential resources that the stakeholder might bring (i.e. on-site visit at their location, training a.s.o., support and opportunities to continue working on the project after the course ends)

2. Not all students experience CBL in similar ways. Some enjoy CBL while some get frustrated. As a solo teacher, it is harder to keep track of students' experience in the course. We want to emphasise the importance of having a teaching team involved in a CBL course, with its members having different roles: lecturer, technical expert, ethics expert, coach, teaching assistant. Make sure that one of the members of the teaching team can observe students' experience and progress in the course to make sure no student is left behind.
3. CBL can lead to immense self-growth and self-awareness. For this, it is important not to decouple the course work from the "self" work. This can be achieved by designing activities that encourage reflection and including contact moments that allow for sharing reflection between peers or with a coach. Students can identify what worked well or less well, what can be changed, but also different aspects related to the process of developing a project or professional development insights.

References

Alves, A. (2019). A tradição alemã do cultivo de si (Bildung) e sua significação histórica. *Educação & Realidade, 44*, e83003

Apel, K.-O. (1998). *Regulative Ideen oder Wahrheitsgeschehen? Zu Gadamers Versuch, die Frage nach den Bedingungen der Möglichkeit gültigen Verstehens zu beantworten.* In: Auseinandersetzungen in Erprobung des transzendentalpragmatischen Ansatzes (pp. 569–607), Frankfurt a. M.

Apel, K.-O. (2011). *Paradigmen der ersten Philosophie*. Suhrkamp.

Backhaus, J. (2015). The university before Humboldt and after. In J. Backhaus (Ed.), *The university according to Humboldt* (Springer briefs in economics) (Vol. 89). Springer. https://doi.org/10.1007/978-3-319-13856-5_1

Bennett, J. (2010). *Vibrant matter: A political ecology of things*. Duke University Press.

Berlin, I. (1976). *Four essays on liberty*. Oxford University Press.

Bertoni, A. (2020). Introducing value driven design in engineering education: teaching the use of value models in preliminary design. *International Journal of Technology Design Education, 30*, 531–552. https://doi.org/10.1007/s10798-019-09511-x

Biesta, G. (2002). *Bildung* and modernity: The future of *Bildung* in a world of difference. *Studies in Philosophy and Education, 21*, 343–351.

Bohlin, H. (2008). *Bildung and moral self-cultivation in higher education: What does it mean and how can it be achieved?* Forum on Public Policy. https://files.eric.ed.gov/fulltext/EJ1099530.pdf

Bombaerts, G. (2020). Upscaling challenge-based learning for humanities in engineering education. In *48th SEFI annual conference 2020 European Society for Engineering Education*, Twente, The Netherlands (pp. 104–114).

Bombaerts, G. (2021). Basic need frustration in motivational redesign of engineering courses. In J. Bennedsen, K. Edstrom, M. S. Gudjonsdottir, I. Saemundsdottir, N. Kuptasthien, J. Roslof, & A. Sripakagorn (Eds.), *17th International CDIO conference, CDIO 2021 - proceedings* (pp. 732–741). Chulalongkorn University.

Bombaerts, G. J. T., Doulougeri, K. I., Spahn, A., Nieveen, N. M., & Pepin, B. E. U. (2018). The course structure dilemma: Striving for engineering students' motivation and deep learning in an ethics and history course. In *46th SEFI annual conference 2018: Creativity, innovation and entrepreneurship for engineering education excellence*.

Bombaerts, G., Doulougeri, K., Tsui, S., Laes, E., Spahn, A., & Martin, D. A. (2021). Engineering students as co-creators in an ethics of technology course. *Science and Engineering Ethics, 27*, 48. https://doi.org/10.1007/s11948-021-00326-5

Bombaerts, G., Martin, D., Watkins, A., & Doulougeri, K. (2022). Reflection to support ethics learning in an interdisciplinary challenge-based learning course. In *EDUCON conference*, Tunis, Tunesia.

Bombaerts, G., & Nickel, P. J. (2017). Feedback for relatedness and competence: Can feedback in blended learning contribute to optimal rigor, basic needs, and motivation? In *IEEE global engineering education conference*. Athens, Greece. https://doi.org/10.1109/educon.2017.7942984.

Bombaerts, G., & Vaessen, B. (2022). Motivational dynamics in basic needs profiles: Toward a person-centered motivation approach in engineering education. *Journal of Engineering Education, 111*(2), 357–375. https://doi.org/10.1002/jee.20448

Brown, T., & Martin, R. (2015). Design for action. *Harvard Business Review, 93*(9), 30–32.

Christensen, J., Henriksen, L. B., & Kolmos, A. (Eds.). (2006). *Engineering science, skills, and Bildung*. Aalborg Universitetsforlag.

Christensen, S. H., Meganck, M., & Delahousse, B. (2007). Introduction. Occupational *Bildung* in engineering education. In H. Christensen et al. (Eds.), *Philosophy in engineering* (pp. 13–22). Academia.

Cordeiro Cruz, C. (2021). Brazilian grassroots engineering: A decolonial approach to engineering education. *European Journal of Engineering Education, 46*, 690.

Crawley, E. F., Malmqvist, J., Ostlund, S., & Brodeur, D. R. (2007). *Rethinking engineering education: The CDIO approach*. Springer.

Diaz Lantada, A. (2020). Engineering education 5.0: Continuously evolving engineering education. *International Journal of Engineering Education., 36*, 1814–1832.

Dilthey, W. (1910/2021). Wilhelm Dilthey: Selected Works, Volume III : The Formation of the Historical World in the Human Sciences. Ed. by R. A. Makkreel and F. Rodi. Princeton: Princeton University Press.

Eck, W. (2019). *Herder's science of man: The origins of anthropology in the philosophy of language*. Doctoral Dissertation, University of Pittsburgh.

Elias, N. (2010). *The civilizing process*. Blackwell.

English, A. (2014). *Discontinuity in learning: Dewey, Herbart, and education as transformation*. Cambridge University Press.

Gadamer, H.-G. (1953). Wahrheit in den Geisteswissenschaften. In H.-G. Gadamer, *Kleine Schriften I: Philosophie, Hermeneutik*, (pp. 37–45). Mohr Siebeck.

Gadamer, H.-G. (1972). Herder und die geschichtliche Welt, in: H-G. Gadamer: *Kleinere Schriften III*, Tübingen, 101–117

Gadamer, H.-G. (1989). *Truth and method (J. Weinsheimer & D. G. Marshall, Trans.)*. 2nd Ed. Continuum.

Gaidi, K. (2009). Engineering education and concept of '*Bildung*'. In *5th International CDIO conference, Singapore Polytechnic, Singapore*.

Gallagher, J. E., & Savage, T. (2020). Challenge-based learning in higher education: An exploratory literature review. *Teaching in Higher Education, 28*, 1135.

Habermas, J. (1981). *Urbanisierung der Heideggerschen Provinz, in: Philosophisch-politische Profile, Frankfurt a.M.* 392–401.

Habermas, J. (1987). *The theory of communicative action (T. McCarthy, Trans.)*. Beacon Press.

Habermas, J. (1991). *The structural transformation of the public sphere: An inquiry into a category of bourgeois society*. MIT Press.

Hadgraft, R. G., & Kolmos, A. (2020). Emerging learning environments in engineering education. *Australasian Journal of Engineering Education, 25*(1), 3–16.

Hamann, J. (2011). '*Bildung*' in German human sciences: The discursive transformation of a concept. *History of the Human Sciences, 24*(5), 48–72. https://doi.org/10.1177/0952695111421865

Heidt, I. (2015). Exploring the historical dimensions of *Bildung* and its metamorphosis in the context of globalization. *L2 Journal, 7*(4), 2–16.

Herbart, J. F. (1802/2022). Introductory lectures to student in pedagogy. In N. Friesen (Ed.), *Tact & the pedagogical relation: An introduction*. Peter Lang. https://blogs.ubc.ca/nfriesen/files/2015/03/Herbart-INTRODUCTORY-LECTURE-TO-STUDENTS-IN-PEDAGOGY.pdf

Herbart, J. F. (1806/1908). *The science of education: Its general principles deduced from its aim and the aesthetic revelation of the world* (H. M. Felkin & E. Felkin, Trans.). D. C. Heath.

Herder, J. G. (2002). Philosophical writings. In M. Forster (Ed.), *Herder: Philosophical writings* (Cambridge texts in the history of philosophy). Cambridge University Press. https://doi.org/10.1017/CBO9781139164634.017

Horlacher, R. (2004). *Bildung* – A construction of a history of philosophy of education. *Studies in Philosophy and Education, 23*(5), 409–426. https://doi.org/10.1007/s11217-004-4452-1

Kauffmann, A. C. J. (2012). *Willhelm von Humboldt and John Stuart Mill on Bildung and self-development*. Thesis submitted to The University of Cambridge.

Kohn Rådberg, K., Lundqvist, U., Malmqvist, J., & Hagvall Svensson, O. (2020). From CDIO to challenge-based learning experiences—Expanding student learning as well as societal impact? *European Journal of Engineering Education, 45*(1), 22–37.

Leydens, J. A., & Lucena, J. C. (2017). *Engineering justice: Transforming engineering education and practice*. John Wiley & Sons.

Martin, D. A., & Bombaerts, G. (2024). What is the structure of a Challenge Based Learning Project? A shortitudinal trajectory analysis of student process behaviours in an interdisciplinary engineering course. *European Journal of Engineering Education, 50*, 51–81.

Martin, D. A., Bombaerts, G., Horst, M., Papageorgiou, K., & Viscusi, G. (2023). Pedagogical orientations and evolving responsibilities of technological universities: A literature review of the history of engineering education. *Science and Engineering Ethics., 29*, 40. https://doi.org/10.1007/s11948-023-00460-2

Martin, D. A., & Conlon, E. (2023). What is engineering ethics education? Exploring how the educa of ethics is defined by engineering instructors. In A. Fritzsche & A. Santa-María (Eds.), *Rethinking technology and engineering: Dialogues across disciplines and geographies* (Philosophy of engineering and technology (POET)) (Vol. 45, pp. 263–282). Springer. https://doi.org/10.1007/978-3-031-25233-4_20

Martin, D. A., Conlon, E., & Bowe, B. (2021a). A multi-level review of engineering ethics education: Towards a socio-technical orientation of engineering education for ethics. *Science and Engineering Ethics, 27*(5), 60. https://doi.org/10.1007/s11948-021-00333-6

Martin, D. A., Conlon, E., & Bowe, B. (2021b). Using case studies in engineering ethics education: The case for immersive scenarios through stakeholder engagement and real-life data. *Australasian Journal of Engineering Education, 26*(1), 47–63. https://doi.org/10.1080/22054952.2021.1914297

Martin, D. A., Gwynne-Evans, A., Kazakova, A., & Zhu, Q. (2023). Developing a global and culturaly inclusive vision of engineering ethics education and research. In A. Johri (Ed.), *International handbook of engineering education research* (pp. 87–114). Routledge Taylor & Francis Group.

Masschelein, J., & Simons, M. (2002). An adequate education in a globalised world? A note on immunisation against being-together. *Journal of Philosophy of Education, 36*(4), 565–584.

Mazzurco, A., & Daniel, S. (2020). Socio-technical thinking of students and practitioners in the context of humanitarian engineering. *Journal of Engineering Education, 109*, 243–261. https://doi.org/10.1002/jee.20307

McGowan, V. C., & Bell, P. (2020). Engineering education as the development of critical sociotechnical literacy. *Science & Education, 29*, 981–1005. https://doi.org/10.1007/s11191-020-00151-5

Mitcham, C. (1994). Engineering design research and social responsibility. In K. Shrader-Frechette (Ed.), *Ethics of scientific research* (pp. 153–168). Rowman & Littlefield.

Mitcham, C., & Munoz, D. (2010). *Humanitarian engineering*. Morgan & Claypool.

National Academy of Engineering (2017). *Engineering Societies and Undergraduate Engineering Education: Proceedings of a Workshop*. Washington DC: The National Academies Press. https://doi.org/10.17226/23686

Niles, S., Contreras, S., Roudbari, S., Kaminsky, J., & Harrison, J. L. (2020). Resisting and assisting engagement with public welfare in engineering education. *Journal of Engineering Education, 109*, 491–507.

Nordenbo, S. E. (2002). *Bildung* and the thinking of *Bildung*. *Journal of Philosophy of Education, 36*, 341–352.

Östling, J. (2018). *Humboldt and the modern German university: An intellectual history*. Lund University Press. https://library.oapen.org/handle/20.500.12657/30499

Pierrakos, O., Beam, T. K., Constantz, J., Johri, A., & Anderson, R. (2009). On the development of a professional identity: Engineering persisters vs engineering switchers. In *39th IEEE frontiers in education conference, San Antonio, USA* (pp. 1–6). https://doi.org/10.1109/FIE.2009.5350571

Pinkard, T. (2010). *German philosophy 1760–1860: The legacy of idealism*. Cambridge University Press.

Platttner, H., Meinel, C., & Leifer, L. (Eds.). (2010). *Design thinking: Understand - improve - apply*. Springer.

Reyes Viviescas, D., Ventura-Medina, E., Anderson, T., & Mio, C. (2019). Development of learner autonomy in student-centred learning environments in engineering. In B. V. Nagy, M. Murphy, H.-M. Järvinen, & A. Kálmán (Eds.), *SEFI 47th annual conference proceedings* (pp. 899–908). European Society for Engineering Education (SEFI). ISBN 978-2-87352-018-2.

Rijk, C. (2019). *TU/e education in 2030*. Eindhoven University of Technology.

Ryan, R. M., & Deci, E. L. (2000). Self-determination theory and the facilitation of intrinsic motivation, social development, and well-being. *American Psychologist, 55*(1), 68–78. https://doi.org/10.1037/110003-066X.55.1.68

Schiefele, U. (1991). Interest, learning, and motivation. *Educational Psychologist, 26*(3–4), 299–323. https://doi.org/10.1080/00461520.1991.9653136

Shaftesbury, A. A. C. (1710/1981). Soliloquy: Or, advice to an author. In G. Hemmerich & W. Benda (Eds.), *Anthony Ashley Cooper, third Earl of Shaftesbury* (pp. 35–301). Fromann-Holzboog.

Spahn, A. (2009). *Hermeneutik zwischen Rationalismus und Traditionalismus: Gadamers Wahrheitsbegriff vor dem Hintergrund zentraler Paradigmen der Hermeneutikgeschichte*. Würzburg: Königshausen & Neumann

Taylor, C. A. (2017). Is a posthumanist *Bildung* possible? Reclaiming the promise of *Bildung* for contemporary higher education. *Higher Education, 74*(3), 419–435. https://doi.org/10.1007/s10734-016-9994-y

TU Eindhoven. (n.d.). *Innovation space*. https://www.tue.nl/en/university/campus/facilities/innovation-space/

Valdemarin, V. T. (2022). The circulation of Johann Friedrich Herbart's ideas in North American textbooks (1893–1920). *Educação e Pesquisa, 48*, e244750. https://doi.org/10.1590/s1678-4634202248244750por

Varkøy, Ø. (2014). *Bildung* - Between cultural heritage and the unknown, instrumentalism and existence. In M. Fleming, L. Bresler, & J. O'Toole (Eds.), *The routledge international handbook of the arts and education*. https://doi.org/10.4324/9781315742717.ch3

von Humboldt, W. (1793/2000). Theory of *Bildung* (Gillian Horton-Krüger, Trans.). In Westbury, I., Hopmann, S., Riquarts, K. (Eds.) *Teaching as a reflective practice: The German Didaktik tradition* (pp. 57–61). Lawrence Erlbaum. https://blogs.ubc.ca/nfriesen/files/2015/01/Humboldt-on-Bildung.pdf

von Humboldt, W. (1850/1993). In J. W. Burrow (Ed.), *The limits of state action*. Liberty Fund.

Yemini, M., Engel, L., & Ben Simon, A. (2023). Place-based education – A systematic review of Literature. *Educational Review, 1–21*. https://doi.org/10.1080/00131911.2023.2177260

Diana Adela Martin Senior Research Fellow based at the Centre for Engineering Education, University College London. She is the co-chair of the SEFI (European Society for Engineering Education) Ethics Special Interest Group, a board member of REEN - Research in Engineering Education Network, as well as an Associate Editor for the journals The European Journal of Engineering Education and Science and Engineering Ethics.

Andreas Spahn Associate professor with the Philosophy & Ethics research group at Eindhoven University of Technology. His research focusses on ethics of technology, particularly ethics of energy technologies. He has worked and published on ethics of behavior change technologies and persuasive technologies, ethics of energy systems, and environmental ethics, as well as on the role of ethics in engineering education. Spahn has been the responsible teacher for a large-scale ethics of technology course, that is mandatory for all first-year students. He is a member of the management team and senior researchers in the 4TU. Ethics of Technology centre, as well as in the ESDiT research program on the ethics of socially disruptive technologies.

Gunter Bombaerts Assistant Professor at the Philosophy and Ethics group at Eindhoven University of Technology (TU/e), the Netherlands. His philosophical work focuses on attention, system thinking and Buddhist ethics. For the last 10 years, he has coordinated the USE (User, Society, Enterprise) program, a 20 ECTS program to make students reflect on the sociotechnical of their engineering discipline. He designed many challenge-based learning (CBL) courses in an interdisciplinary way. In order to improve the quality of this program, he started performing research in engineering ethics education, focusing on motivation, self-regulation, and student activation. He also studies how pedagogies such as CBL impact the university's organisation, epistemologies, and role in its ecosystem.

Chapter 6
Thinking About Engineering Education: Between *Bildung* and Citizenship

François Prouteau, Nathanaël Wallenhorst, and Renaud Hétier

Abstract This chapter sketches an emerging orientation in engineering education, by proposing the reciprocal articulation of the concepts of *Bildung* and of citizenship through the paradigm of "existential citizenship." After introducing the original notion of *Bildung*, rooted in nineteenth-century German culture, we propose its integration with the concept of citizenship. The latter helps us transcend individual-focused readings of education in favor of the uptake of responsibility for the world, understood in the arendtian sense of political action. The result is the concept of existential citizenship, which we offer as a guide for thinking about engineering education; the existential dimension of this concept focuses attention on the education for humanity inherent in the notion of *Bildung*, whilst retaining the political dynamic of investment in the transformation of the city.

Keywords *Bildung* · Citizenship · Political action · Existential citizenship · Engineering education

---

The authors would like to thank Luigi Russi, PhD, for English language editing.

F. Prouteau (✉)
Université Catholique de Lille, Lille, France
e-mail: francois.prouteau@univ-catholille.fr

N. Wallenhorst
Faculty of Education, Universite' catholique de L'Ouest, Beaucouze, France
e-mail: nathanael.wallenhorst@uco.fr

R. Hétier
Universite' catholique de L'Ouest, Angers, France
e-mail: Renaud.hetier@uco.fr

A. Buch, S. H. Christensen (eds.), *Bildung for Engineering Education and Practice*, Philosophy of Engineering and Technology 49,
https://doi.org/10.1007/978-3-031-86581-7_6

6.1 Introduction

Pour un réveil écologique [Ecological Awakening] is a contemporary movement within French higher education, which was spearheaded by graduates of the country's elite engineering schools. Its manifesto, signed by over 30,000 students—among whom a preponderance of engineering students—illustrates how actions and words have the power to change political discourse, in accordance with a classic lesson learned in the Humanities: "the conviction that these two human capacities belong together and are the highest of all abilities seems to have preceded the *polis* and was already present in pre-Socratic thinking" (Arendt, 2018, p. 25). *Pour un réveil écologique* has had, and continues to have, a real impact on economic and political decision-makers, and also on higher education itself, as the movement produces practical guides and tools, notably to assess the content and context of educational offerings around ecological transition at engineering schools.

The rise of a movement like this, and the fact that it takes responsibility for steering the curriculum of engineering schools, challenges education scholars to conceptualize an emerging dimension that is surfacing in education in general, and in engineering education in particular: the education of citizens. When we speak of education[1] of citizens in higher education, we must accept a three-faceted normative idea: firstly, that the mission of academic institutions is not only that of professional training, but also more broadly educational (allowing students to learn about the world); secondly, that a citizen has the right to belong to the City in educational establishments set up to cater for the members of a State, and which have a genuinely political role; and thirdly, that a citizen's own self-education and the education system as a whole are not axiologically neutral, but function in reference to a particular political ideology. This last point is fundamental for making sense of the place of contemporary movements, like *Pour un réveil écologique*, in the context of engineering education: it brings home the reality that thinking about the education of citizens in engineering schools cannot be divorced from asking challenging questions concerning available representations of "the good life"—something that stresses the political nature of education (Ardoino, 1999).

The scholarship of Franco-German economist Christian Arnsperger provides helpful resources to conceptualize this emerging dimension. Arnsperger's work introduces the term "existential citizens"—citizens who make radical life choices, and voluntarily break with their so-called "capitalist existence" (2010, 2011a, 2011b). The advantage of this notion is that it allows us to view an existential component in the context of political education, and arrange it so that it lays the foundations for the subject's active participation in the world (Arendt, 2018). This comes at a time when, in engineering schools, Transition (ecological, energy and societal

[1] The idea of adult education emerged during the twentieth Century and has, historically, been institutionalized outside of the University. We have seen a semantic shift since the emergence of "adult education" or "permanent education" toward "continuing development" (though the notion of "lifelong education" has also been around for the past 15 years or so).

transition) is becoming a major priority in the education provided to students, with input from political and economic decision-makers.

Accepting the challenge that movements like *Pour un réveil écologique* raise for engineering education, this chapter is an attempt to articulate it conceptually through a combination of the traditional notion of *Bildung* with the idea of citizenship. *Bildung* is a notion that grew in popularity and reached its peak in nineteenth-century Germany. It is endowed with a unique and crucial function, bringing together the Universities, and mediating among individuals, the community and the State. Thus, *Bildung* has historically embodied the aspiration for ideal education and shaping of human subjects, but also fulfilled a social role (Masschelein, 2002, pp. 37 and 45). In Germany, the role of *Bildung* was at the very core of the foundation of the Humboldtian University, whose mission was more broadly educational, in addition to that of strict and formal training.

The idea of *Bildung* is characterized by the ambition to achieve an overarching education, which shapes a subject's very humanity, beyond their skills and adaptation to the requirements of society—and, more specifically, to the world of work. The question then arises of its relevance in relation to the education of engineers, which is, *a priori*, marked by the quest for scientific knowledge and technical skill, as well as operational and material efficiency. Here, a variety of possible configurations seem possible, depending on the approach one chooses to take. At one extreme, engineering education can be envisioned through a prism of pure technical skill based on scientific excellence, which would merely be encumbered by complex and sometimes contradictory human developmental dimensions. At the other extreme, our preoccupation is with humanity, as we begin to suspect—as Heidegger did—that all technology may have an alienating effect.

Some scholars (Sjöström et al., 2017) have approached this alternative by distinguishing up to five different notions of *Bildung*, further arranged into three subgroups. A first sub-group comprises (1.a) German *Bildung* (in its original nineteenth-century understanding) and (1.b) Anglo-American liberal education. A second subgroup consists of (2.a) Scandinavian notions of *folk-Bildung* and (2.b) democratic education rooted in American pragmatism. The last subgroup encompasses (3) critical-hermeneutic *Bildung* fashioned through the lens of Hans-Georg Gadamer's hermeneutic philosophy. In broad strokes, Sjöström et al. (2017) suggest that the first sub-group centres on individual skill development and personal growth, the second on preparation as functioning members of society, and the third on responsible citizenship, emphasizing a capacity for political mobilization around emerging issues like social and ecological justice. Our proposal in this chapter charts a similar progression, but it does so through a different route. This is because the term *Bildung* does not have a straightforward translation into languages like French, English or Italian (Liedman, 1997, quoted in Sjöstrom et al., 2017, p. 166): it cannot easily be translated into such languages, but it generally refers to education, learning or culture "of the self."[2] Therefore, some of the developments that Sjöström et al. (2017) situate within the semantic field of *Bildung* have not been associated with this notion everywhere. Such is the case of France, for example, where it is rather discourse around *citizenship* that gathers together educational philosophies similar to those that Sjöström et al. (2017) prefer to associate with *Bildung* (subgroups two and three above). Indeed, citizenship can accommodate both technical ability (employed to benefit a city wishing for development, progress and continual improvement) and the human concern that *Bildung* embodies.

Our proposal, then, is to articulate together the spirit of *Bildung* and the practicality of technicity, under a unified approach of citizenship. This chapter, which belongs to the field of political anthropology, can therefore be read as a parallel endeavor to that of Sjöström et al. (2017): in our attempt to make sense of recent trends in engineering education, we notice—like those authors also do—an emerging shift from a focus on the individual or even on the competent member of society towards a responsible citizen, who is able and willing to call into question the implicit representations of what a "good life" ought to be. In our reading, the concept of citizenship provides an additional basis for reflecting on engineering education: this interweaving of notions allows us to understand political responsibility for the world in relation to others, and to understand the intimate nature of our own existence. By making the connection between them, our goal is to enrich the concept of citizenship with existential resonance, through the notion of "existential citizenship." In order to do so, we pursue a conceptual dialogue between the notion of *Bildung* and that of citizenship, which will have the advantage of speaking to readers who might be less familiar with the discourse around *Bildung* for linguistic and/or cultural reasons.[3] Finally, our articulation of *Bildung* and citizenship takes place primarily at a paradigmatic, theoretical and conceptual level, whereas the empirical component is not explicitly discussed.

This being our general intention, the chapter pursues it in the following steps. The first section discusses the place of education and politics in the context of engineering. The second section identifies the strengths and limitations of the concept of *Bildung* (in its original nineteenth century formulation), when it is read as a political conception of education, and specifically with reference to engineering education. The third section introduces the notion of citizenship, once again in relation to engineering education, by offering an overview of the distinction between labor, work, and action (Arendt, 1961, p. 62)—with their reciprocal articulation being the proper domain of citizen's political concern and responsibility. Section 6.4 brings together the original Germanic perspective on *Bildung* with the notion of citizenship—conceived of as a paradigm of education for politics (Ardoino, 1999; Murard & Tassin,

[2]The notion of *Bildung* is generally described by three propositions: "self-education", "self-training", and "self-culture". It is preferable to refine this three-faceted idea by highlighting the self, which is the true object of *Bildung*.

[3]This does not detract from the possibility of reaching similar conclusions as we do here, by formulating an emerging critical-hermeneutic notion of *Bildung*, which is what Sjöström et al. (2017) have chosen to do.

2006; Tassin, 2004). Finally, we locate this intersection of *Bildung* with citizenship in Arnsperger's concept of "existential citizenship," which helps orient how, in today's world, we might think about engineers' political education.

6.2 Education and Politics

There are two distinct groups (or poles) of skills that need to be mastered in engineering education. One aims more forthrightly at professionalization (endowing engineers with scientific and technical knowledge and professional attitudes). The other is focused on cross-cutting skills of a political, relational, and ethical nature. This second cluster of skills is related to what Arendt calls "love of the world" and "assuming responsibility for the continuation of the world", which she believed to be the ultimate objective of education (Arendt, 1961, p. 196). In our discussion here, this second pole is viewed as being of particular importance, and we will discuss its philosophical underpinnings in greater depth in Sect. 6.3. It relates not to professional abilities in the strictest sense, but to seminal skills on which professional or technical abilities can be built. Learning about the world is a fundamental capacity for engineers who, through their creations, help to shape that world. For Arendt, this implies the capacity to take care of the world—where "world" means the environment we all share, which is also the space that divides all humans—and pay attention to what is sustainable. It involves the ability to take up responsibility for the world with respect to others. Arendt attaches enormous importance to the world, as the world is the very thing that allows for "natality."

In Arendt's view, to teach students about the world is the educational function of school, which she distinguishes from higher education, since the latter is focused more on professional training. However, the higher education system has more than the simple function of professional training; it is also in charge of carrying forward that so elusive—and yet so indispensable—fundamental education that, even during the 50s, Arendt already identified as being in crisis (her essay "*The Crisis in Education*" was first published first in 1958, then again in *Between Past and Future*). Higher education must continue to guide students in their integration into adult life, taking on responsibility for the world with respect to others and to generations to come (Jonas, 1990, 1998). This requires the extension of educational action, which, Fichte believes, has the direct function of enabling Man to become human (Lamarre, 2002b).[4] Man becomes human through education—he is *homo educandus*—and through politics, it being understood that the objective of education is political, in the sense that it includes participation in the world and assumption of responsibility

---

[4] This study follows in the footsteps of the work of these two German authors, shaped by the notional universe of Germany and their breakthrough humanistic ideas, aiming to perfect the humanity in humans: Fichte, with his emphasis on education, and Arendt, for whom Man is defined by politics. However, it should be noted that Arendt's thinking represents a break with German humanist idealism.

for the world. In the Greek idea of *paideia*, education has the function of allowing citizens to become better human beings: "You have been instructed in these activities not in order to acquire technical ability, but for your education, as is fitting for a simple individual and a free man" (Dialogue, Socrates to Hippocrates, Platon, 1997, p. 72).

In view of this, we place engineering education today at the crossroads of a "professional" approach (the mastering a technique) and a "non-professional" one. How is it possible, by means of professionalization, also to facilitate the assumption of political responsibility with respect to the world and to others? Professional education, including mastery of "best practices," is complementary to taking a position in relation to the world. Engineering education has always adapted to the challenges posed by progress in science and technology, and also to the demand for progress and "acceleration" in modern societies (Rosa, 2013). Today, it must shift position once more, to take account of the socio-ecological crises and risks that threaten our planet's very habitability. Increasingly, this issue is calling into question the place occupied by environmental concerns and the societal quest for peace and justice in twenty-first-century engineering programs. In recent years, there have been multiple publications on this issue, in the form of manifestos which have gained great traction in engineering schools in France, such as the *Manifeste pour un réveil écologique* [Manifesto for Ecological Awakening], published in 2018. Documents such as this highlight the need for major political mobilization of higher education as a complex system made up of various institutions, all laboring in one way or another to bring about a change in the world: States and regional authorities, intergovernmental institutions; educational institutions; students and alumni coming together in various associations; all types of businesses; other stakeholders such as in the media, digital technologies, etc. Another influential document, the *Manifeste pour l'évolution de la formation des ingénieurs* [Manifesto for the evolution of engineering education] (2022), developed by the *Shift project*,[5] calls for the development of innovative and active pedagogical approaches, suitable for addressing these issues, leading engineers to become people who take care of the world. We recall the maxim "*primum non nocere*" (First, do no harm), so dear to Hippocrates (in 410 BCE). This principle has always been a point of reference for doctors and caregivers in their professional practices, and it should also be applicable to engineers, in the ways in which they invent, design and produce.

[5] https://theshiftproject.org/wp-content/uploads/2022/06/Manifeste-Climatsup-INSA-version-Web.pdf

6.3 *Bildung* as Whole-Person Education

The German intellectual tradition of sciences of mind (*Geisteswissenschaften*) offers valuable resources for thinking about engineering education. The conceptual resources from the world of *Bildung* enable us to think of engineering education as a complete, overarching process, which includes reflection on the engineer's humanity. Indeed, in the German tradition of *Bildung*, education and training are explicitly designed and delivered on the basis of existential experience. This approach has the advantage of educating people who are aware of the fundamental questions of humanity, capable of recognizing other beings as *alter ego*. Having been awakened to their own humanity, students formed in the *Bildung* tradition are capable of joining with others and sharing their common humanity. One of the benefits of the notion of *Bildung* lies in its non-professionalizing, non-technical component. It provides an existential foundation for the professional skills of work on behalf of others.

Generally, this notion of *Bildung* has been positively received even outside of Germany for two main reasons: its reference to a whole-person approach, and the fact that its learning takes place beyond the strict framework of the classroom, with a spotlight on the way in which "life" can teach us lessons through our experiences (Brougère & Ulmann, 2009). Amongst the strengths of *Bildung* is the understanding of the subject through a holistic anthropology that favors integration of the directly existential component whereby *Erfahrung* (experience) is *Bildung*; existential experience is viewed as being formative. This concept resonates, for example, with the theory of anthropoformation put forward in France by Pineau (2003). In the French literature on *Bildung*, mention is frequently made of the idealistic universe[6] in which the concept was born, and of the ideals that it carries to this day. However, there is a nuance that has sometimes been lost in the reception of the concept of *Bildung* outside of Germany, and particularly in a country like France, where it lacks an adequate translation. This concerns the political dimension of that idealism. The idealism of *Bildung* must be viewed in connection with the political sphere, as well as to the educational sphere. *Bildung* is a political concept in that it views civil society as being formed by the very process by which the subjects come into their own. One of the virtues of *Bildung* is that social connection is viewed as being a pedagogical matter (Masschelein, 2002, p. 40). The notion of *Bildung* includes the idea that it is possible for Man to create a new Man.[7]

---

[6] In the German world, Herbart sounded a different note. Whilst attaching crucial importance to *Bildung* and to the development of "the humanity in Man through education" (Maigné, 2002, p. 51), he developed a realistic and empirical approach to this notion, which broke away from its idealism. On the other hand, his work was shaped by the individualism which underlies the notion of *Bildung*, in which the very objective is the realization of Man as such.

[7] Perfectibility is an essential concept in pedagogical anthropology of the German Enlightenment, which is constructed around the notion of *Bildung* (Ricken, 2002). Then, as modernity came about, Germany saw the emergence of a critique of this idealistic philosophical anthropology, with the historical anthropology of the Berlin School, and thinkers such as Wulf and Gebauer.

There are also limits to how far the original nineteenth-century notion of *Bildung* can go, of which two particularly stand out and appear indissociable from one another: structural individualism (Dumont, 1991) and a lack of education for politics. In the universe of *Bildung*, the act of educating is viewed as a political act, contributing to the emergence of a new type of human being. Its idealism notwithstanding, one of the strengths of *Bildung* is its ambition. On the other hand, paradoxically, the notion of *Bildung* refers to politics in a highly individualistic way, because politics is designed on the self, based on the self, and created for one's own self. The idea of collective belonging holds little weight in this universe. The subject is part of a world, which is viewed as the arena for self-development. However, the realization of the subject takes place within themselves. It is in the intimate sphere of existence that self-development takes place, and this is the primary goal of *Bildung*. With *Bildung*, we are dealing with self-education, which the subject discovers and develops through contact with the world. Historically, this notion of *Bildung* emerged in an environment of pietistic Protestantism in the late eighteenth century in Germany. Protestantism, with its emphasis on individual faith (to the detriment of institutional authority), reinforced the importance of the inner self, which became the entrance way to the outside world in which the believer must prove themselves (Graf, 2002). With Protestantism, politics is grounded in the intimate experience of existence. *Bildung* occurs, first and foremost, in a person's inner being (Graf, 2002, p. 783), rather than out in the world, through confrontation with others who inhabit it. Individuals are obliged to work on their personal realization (due to the intimisation of salvation), rather than to address the wants or needs of the world.

This tension between inner personal development and involvement in the world, which can be found from the start in the idea of *Bildung*, gestures towards a political view of education with existential foundations, albeit without going as far as contemplating true political education. While, in the spirit of *Bildung*, educating remains a strongly political act, essential for humanity, it is not so much a citizen who is *gebildet* (educated, trained, cultivated), but a human being. The man who receives training is essentialized—it is that man's humanity that is educated or elevated, rather than a citizen or politician being created, anchored in a given economic-social reality. From an idealistic perspective, by elevating the humanity in everyone, we work to create a new human adventure in whose hands the world has a future, without having resorted to political education of the subject as a citizen.

To conclude, *Bildung* contains the seed of a political conception of education with existential underpinnings, albeit without truly integrating education for politics. Nevertheless, it is plainly distinct from the concept of training, which is often limited to the assimilation of learning objectives and acquisition of technical skills, to which numerous engineering "education" programs have often been reduced. The political implications of the original notion of *Bildung* also reveal other limits in relation to the phenomenon of resonance—as elicited by the educational process—when it considered from a political point of view. In resonance theory, the primary objective is not cultivation of the self. Education in the sense of resonance theory is aimed at "cultivating relationships to the world. The goal is not refinement of the individualistic or atomistic self, nor disengaged mastery of the world, but rather opening up and establishing axes of resonance" (Rosa, p. 241). Rosa goes further with his analysis, showing how the educational vision of authors such as Humboldt prefigures school as being a space for resonance: in the teacher/student/material triangle, setting the axes of resonance into vibration, so that a transformative assimilation takes place. By cultivating a transformative experience of the subject who allow themselves to be touched by the world, resonance also encourages the use of scientific methods that develop an approach that is more sensitive to the physical and social environment, and a reflection on the impact of the activities of scientists and engineers on the evolution of the world, as we are encouraged to do by what we now know about the Anthropocene.

The original sense of the term *Bildung* makes limited reference to the political dimension of belonging to a collective. This intrinsic limit is one that some commentators have sought to address by putting forth "updated" versions of the original concept, such as a notion of "critical-hermeneutic" *Bildung* (Sjöström et al., 2017). This will not be our strategy here. In its stead, what we are more interested in doing is intersecting the original, nineteenth-century spirit of *Bildung* with the notion of citizenship, given that the latter enjoys wide currency, including in cultural contexts where *Bildung* is still primarily identified with the original nineteenth-century German version.

When we speak of citizenship, we must inevitably speak both of politics and of education *for* politics. The concept of citizenship, etymologically, refers to the Greek notion of the City, and participation in the life of the City. By comparison with *Bildung*, citizenship has another anchorage point in the outside universe, which is the world. With *Bildung*, the subject is considered as a whole, thought of partly as "self-contained" and marked by a form of cut off from the world, whereas the notion of citizenship is anthropologically more definite (in part, normed and practice-centered); the subject is thought of in terms of their relationship with the world and participation within it. The existential component is thought of in relation to the notion of citizenship, which is connected to politics.

Even if *Bildung* is not a concept from which French higher education refers, this approach is often referred to in terms of "formation intégrale" more and more included in interdisciplinary learning, critical thinking and complex solving, ethics and philosophy courses, humanities and social sciences to deepen their understanding of societal issues and also, community engagement and social responsibility. For example, the *Manifeste pour l'évolution de la formation des ingénieurs* [Manifesto for the evolution of engineering education] (2022) developed by the *Shift project* takes a three-pronged approach: to train engineers who are aware of, and able to understand and analyze, the growing complexity of the world in which they work; to respond to the growing demand for meaning on the part of students, both in their training and in their future professional careers; and to meet the current and future skills needs of companies from a socio-ecological point of view. Such an approach requires more than a proliferation of 'climate entries' in the curricula, as is still often the case in many French Grandes écoles, but an in-depth overhaul of the teaching, which is still rarely undertaken with a systemic and complex dimension.

The Competency-Based Approach (Approche Par Compétences, APC) is a methodology that more and more higher schools in France are using to build their curricula, following the recommendations of the 2022 working group led by Jean Jouzel, a report entitled 'Raising awareness and providing training on the challenges of the ecological transition and sustainable development in higher education'.[8] Including ethics, humanities, human and social sciences, sustainable development and CSR varies from one institution to another and is often laborious to implement; it comes up against well-established cultural habits or silos between disciplines or departments. The competency-based approach combines theoretical knowledge with the ability to act, based on actions to achieve ecological transformation, with general reference to the 17 goals (SDGs) adopted by the UN. The way in which ethics, human and social sciences, sustainable development and CSR are included varies qualitatively and quantitatively from one institution to another,[9] and is often slow and laborious to implement; it comes up against entrenched cultural habits or silos between disciplines or departments, and also to the technicist rationality in which engineering training has certainly developed since the industrial revolution, but has become polarized and confined. From this point of view, the place of the Humanities in the curricula of the engineering schools of the Catholic University of Lille, for example, is a key issue. It focuses on the meaning of the reference to the humanism in which *Bildung* is rooted, both education and culture for learning to assert oneself as a subject in a common model of humanity and of a singular existence narrated by a work of reflexivity and the action of a responsible agent as "being among humans": *Inter homines esse* is the motto of the political and speaking animal (Arendt, 2018).

6.4 Citizenship as Learning About the World and Taking Responsibility for It

Having proposed an integration of the original spirit of *Bildung* with the notion of citizenship, it is helpful to show what we might mean by the exercise of citizenship. This leads us to consider the political theory developed by Hannah Arendt in *The Human Condition* (1958): to what extent can it shed light on the political education of engineers today? Throughout that book, Arendt offers a profound existential

[8] https://www.enseignementsup-recherche.gouv.fr/fr/remise-du-rapport-sensibiliser-et-former-aux-enjeux-de-la-transition-ecologique-et-du-developpement-83903

[9] The Ecological Transition Barometer, inspired by the Manifesto for Ecological Awareness, has carried out a national survey of the extent to which ecology is taken into account in higher education in France. of more than twenty engineering schools, drawing up an inventory of the integration of these issues (https://pour-un-reveil-ecologique.org/fr/grand-barometre/#reponses). Another example is provided by the Grandes Ecoles and Faculties of the Catholic University of Lille, which have established and want to implement Humanities at the heart of digital, technological, economic, ecological and social transitions (https://www.univ-catholille.fr/wp-content/uploads/2024/07/plaquette-DU-Humanites-2024.pdf)

reflection on the human condition in the modern world. She questions the training of all those who, like engineers, will enterprise and innovate to promote scientific and technical progress to benefit society. It should be noted that her book makes no reference to *Bildung*, though educational issues are mentioned multiple times, in the context of various topics. This may be a hallmark of the Arendtian critique of *Bildung*, which can be found explicitly in one of the philosopher's other works, published around the same time—*Rahel Varnhagen: The Life of a Jewish Woman* (Arendt, 1974). While describing the life of an important figure in German Romanticism, Arendt displays profound skepticism towards the idea of *Bildung*: she describes a German Jewish woman who adopted the ideal of self-culture, "only interested in *Bildung* to compensate for the lack of social definition, for being what Hannah Arendt so incisively termed 'pariahs'" (Motzkin, 2001). It is an example of an individualistic or collectivist approach to *Bildung*, which does not encompass the broader field of education that takes account of the plurality needed in order to live well together in the modern world—which is the subject of *The Human Condition*.

Arendt's study of the human condition pertains, firstly, to the systemic transformations of humans' experience and living conditions, brought about by technical progress and modern epistemology, spearheaded by Galileo and Descartes: these have profoundly influenced modern conceptions of the sciences and how they should be taught, even up to the present day—and, of course, this holds true also for engineer education. Hannah Arendt, with perspicacity that also endures to this day, as we have eventually entered the Anthropocene epoch, also sets out to promote a political perspective for making the world a better place, in relation to the new powers conferred upon humanity by scientific progress. It would do so, firstly, thanks to all the beneficial contributions those powers make to our lives, and secondly, by helping to better our understanding of the harmful effects and serious threats that this same technical progress can bring about. It is interesting for our purposes to examine the critique of human activity, which is at the heart of Arendt's political theory, based on three fundamental components—*work*, *labor* and *action*—which are very frequently confused. By distinguishing and analyzing each of them, we can reflect more deeply on the political education of engineers in relation to the notion of citizenship.

6.4.1 Work

The word "*work*," in the context of Arendt's writings, refers to creative efforts to produce physical objects, which give form to the world, and allow us to connect with other people and to make progress. At first glance, *work* is indubitably the component of activity, which is most closely related to the skillset of engineers. After all, is not an engineer's purpose to invent machines to improve living conditions and optimize the impact of our activities? From that perspective, engineers' activity is devoted to innovation, to the creation or manufacture of objects in a race for progress. The history of progress from *homo faber* to the contemporary human adventure could be appraised by the results of ingenious research which, while demanding a "*little work*" of geophysical forces, produced a huge effect, to the point of shifting the Earth.

From this point of view, Hannah Arendt (2018, p. 305) refers to the ancient statement attributed to the genius, Archimedes: "Give me a lever long enough and a fulcrum on which to place it, and I shall move the world [or, according to an oft-used but less exact translation, I shall lift the world].[10]" Should this lead Man to the point of breaking free of the limits and forsaking "the Earth, which was the Mother of all living creatures under the heavens'"(Arendt, 2018, p. 2)? We cannot fail to notice the relevance of such a question today, when we look at the fact of the Anthropocene: in recent decades, scientific research has proved the massive impact of human activities on the trajectory of the Earth system, since the First Industrial Revolution, with massive acceleration since the mid-twentieth Century.

This fulcrum of which Archimedes spoke, according to Arendt, was discovered at the dawn of the Modern Age, not outside of Earth's expanse, but in what it has produced with better performance in *homo sapiens* and *homo faber*: the techno-scientific knowledge so dear to engineers. The genius of Galileo marked an important tipping point in this history, when he invented the telescope, which represented the Holy Grail in the quest to find Archimedes' fulcrum. By contrast to the scientists who went before him, Galileo learned to have a different view of reality: an instrument can produce measurements and data that are more reliable than those obtained by direct observation with human senses. Thus, a change took place in humanity's relationship to nature, with increasing importance accorded to mathematics, linked to measuring instruments, and to the universal dimension of the laws thought to govern the order of all things. With a view to discovering universal laws, experimentation on the basis of human understanding and of the new possibilities offered by technology then gained favor over the observation of natural phenomena. According to Hannah Arendt, there was first a separation between Being and Appearing, in the sense that contemplation is no longer sufficient to reveal truth. One must also beware of deceptive appearances and create objects to verify our observations. Ultimately, the process of manufacture and consumption of objects has increasingly become the norm, to the point where this process actually precedes the object itself, leading to a "reversal of ends and means" (Arendt, 2018, p. 297).

What becomes of the human condition when "the productive and limited instrumentality of fabrication changes into limitless instrumentalization of everything in existence" (Arendt, 2018, p. 157)? We may legitimately wonder about the fate of a world that is no longer civilized, but is instead vulgar or barbarous—i.e. a world where, in the activities of those who invent these artificial objects, there is no active love for beauty, either in the political domain or in education, with people only pursuing their own glory, won by means of what they produce. "Only the vulgar will condescend to derive their pride from what they have done; they will, by this

[10] Pappus of Alexandria, *La Collection Mathématique*, translated into French by P. Ver Eecke, 195, vol. II, p. 836.

condescension, become the 'slaves and prisoners' of their own faculties and will find out, should anything more be left in them than sheer stupid vanity" (Arendt, 2018, p. 211). Arendt's remark may serve as a warning to all professionals—in particular, scientists and technicians who innovate and create new objects—about the idolatry of *homo faber* when one believes that what one produces is more important and more essential than oneself. In sum, in her discussion of work, Arendt exposes the risk implicit in confusing "education" with mere training in a narrow set of technical skills.

6.4.2 Labor and Action

A second component of human activity, in Arendt's analysis, is *labor. Labor* is what we do to serve our biological needs, to keep ourselves alive, or propagate human life through sexual activity. Arendt, after recapping the distinctions between *work* and *labor* proposed by other philosophers before her—notably Aristotle and Locke (Arendt, 2018, p. 80)—stresses that, since ancient times, *labor* has been associated with those who have been deprived of the right to take part in political functions. Such people carried out servile tasks to serve the needs of sustaining life (non-citizens—slaves, in particular), whereas citizens abstained from such labor (*skholè*), focusing exclusively on politics,[11] education and philosophizing (Arendt, p. 81). In contrast to this, what Marx showed is the need to establish philosophy, political thinking and education that are as close as possible to the actual reality, reflecting the vital issues for the existence of all humans, their original state of precarity and dependence which unites everyone in a common condition of interdependent labor. *Labor*, the activity of production, is the birthplace of truth (Vioulac, 2018): in materiality, in contact with the Earth and with the living world. In the face of the challenges posed by the Anthropocene, engineering schools today are realizing that the living world needs to be brought back to the heart of the curriculum and they are striving to imagine how to concretize this new awareness of the importance of the living world for engineers' ways of living and their future employment.[12]

However, this shift at the level of engineering curricula must, more broadly, embrace education for integral ecology, which also addresses environmental, ethical, and social questions. In fact, the Anthropocene demonstrates the pattern of

[11] Arendt, p. 81: "Contempt for laboring originally arising out of a passionate striving for freedom from necessity and a no less passionate impatience with every effort that left no trace, no monument, no great work worthy of remembrance, spread with the increasing demands of polis life upon the time of the citizens and its insistence on their abstention from all but political activities, until it covered everything that demanded an effort [...] The opinion that labor and work were despised in antiquity because only slaves were engaged in them is a prejudice of modern historians. The ancients reasoned the other way around and felt it necessary to possess slaves because of the slavish nature of all occupations that served the needs for the maintenance of life".

[12] "L'Ingénieur et le Vivant", third Forum across a dozen campi of the Institut Mines-Télécom, 19 January 2023.

domination of our society, which has led to the violent exploitation of nature and of the *homo laborans*. Karl Marx, in his critique of capitalism, studied the bases of this pattern of domination. His critique was ontological, as Jean Vioulac has shown, because it analyses the reversal of the object and the subject, by the growing domination of the former over the latter: humans are no longer subjects—they have become the objects of knowledge in scientific objectivism. They have been robbed of their being, to serve the production of objects that are increasingly powerful and coveted; exploited and distanced from the conditions of existence for all Earthlings; subjects are extenuated, alienated, and reified. For this reason, if the question of fairness arises, it is in the sense of restoring all people to their existence (Vioulac, 2022, p. 224). Therefore, paying closer attention, in engineering education, to the reality of the world in connection with the conditions of life exemplifies education as learning about the world (Arendt, 1961) and constitutes a pre-requisite for asking fundamental questions that invite the uptake of responsibility—this being the proper realm of political action by citizens.

Indeed, action, according to Arendt, is our singular manner of making our mark on a multifaceted world; of beginning something new as a person, firstly through birth, and then in everything that we do, that we have done and that we want to do. Thus, action allows us not only to construct stories that we can tell, but also, within the diverse human group, reveals the agent in actions and words (Arendt, 2018, p. 175). When considered through the prism of action, engineering education still falls short in many respects. It falls short every time it separates out, on the one hand, the questioning of suicidal trajectories in our productive systems or the reconsideration of industrial processes as well as of financial, legal, social and human architectures (Travadel & Guarnieri, 2021); from—on the other hand—the ways in which an engineer might become engaged as an actor capable of speech and action, as an author capable of exercising creativity and of accounting for his/her deeds in the first person. When we take this rather striking separation into consideration, we become better able to appreciate the originality of movements like *Pour un réveil écologique*. Movements like this one foreground, through their actions and in their manifestos, the uptake of responsibility and the willingness to "take one's future in one's hands." As highlighted previously, the inspiration for such movements appears to be the capacity of mobilizing "subjects" through dialogue, reflection, and associated projects involving the range of stakeholders concerned by the transformation of education in light of ecological transition—something which demands at one and the same time competent engineers *and* competent citizens. This offers a telling example of how citizenship can play out in engineering education by encouraging the assumption of responsibility—a point to which we turn in the next section.

6.5 Existential Citizenship at the Intersection of *Bildung* and Citizenship

At last, we arrive at the core of our proposal: to integrate the spirit of *Bildung* with the notion of citizenship, which is more directly political (in the sense of Arendt's idea of learning about, and taking responsibility for, the world). By thinking of engineering education on the basis of these combined notions, we are able to grasp political responsibility for the world in relation to other people (implied by the concept of citizenship), whilst retaining a connection to the intimacy of our own existence (implied by the original sense of *Bildung*). This approach invites to join the noun "citizenship" to the adjective "existential", in such a way as to evoke the richness of the notional universe of *Bildung* and the importance of its existential component. The strength of a notion of "existential citizenship" is the ability to apprehend the existential component in the context of political education, with a view to being able to participate in the world. From this perspective, the subject's existence is understood as a possible point of support to allow them to open up to the world. Hence, thinking of engineering education on the basis of the concept of existential citizenship has two implications: (1) the importance of subjects' existential learning based on their relationship to the world and their participation in it; and (2) the political responsibility of engineers in mobilizing their capacities to diagnose the transformation of human experience not only by means of technology, but also through the connection between sense experience and thought (Arendt, 2018, p. 287)—relying on the complex relationship between action, work, and labor to influence what they do.

The expression "existential citizenship" that we adopt here was originally coined by the Franco-German economist Christian Arnsperger. He worked on that notion 'Economic Man and the Meaning of Life' (2011) and, in a way, education for existential citizenship underpins his entire work. The notion of existential citizenship shows a degree of connection with that of *Bildung*, as demonstrated by a statement such as this: "we must highlight the dimension of *human existence as experimentation, as a search for the self, and freedom*" (Arnsperger, 2011a, p. 76). In his work, we find the characteristic traits of that value (*Bildung*) and its experience of existence. Existence is thought of as the predominant experience upon which one's actions should be based. On the other hand, existential citizenship also adds something more to the original sense of *Bildung*, as understood in 19th-centry Germany: existential citizenship has an entirely different goal than simply engendering awareness of one's existence and finding one's own freedom.

Arnsperger's concept of existential citizenship evokes at once (1) a pronounced capability for individual reflection on the contemporary period (in particular, the alienating hegemony of the capitalist system) and (2) an awareness of the existential component of the human adventure. Existential citizenship lies at the boundary between the position of the individual in the contemporary world and their responsibility in terms of laying the foundations for the future, and of questioning the meaning of life: "If the citizen of tomorrow fails to become more aware of the existential roots of our system, of their own relationship with death anxiety and the dread of poverty, we shall continue merely to exist (…) within the ever-narrower margins that social-democratic capitalism affords us" (Arnsperger, 2011a, p. 75). Enabling students to act as citizens by taking up responsibility ought undoubtedly to be one of the major aims of political education under the paradigm of "existential citizenship." Paul Ricœur speaks of such a goal being the inspiration behind the arendtian phenomenology of disclosure of who, the revelation of agents in words and deeds; in the public domain and in the web of human relations, their labor, work and action together constitute political life. Ricœur, who wrote the foreword to the French translation of *The Human Condition*, bases his analysis on the observation of the real world and of human phenomena in society, and then works backwards to the "ability to do" which gave rise to these observable facts: the ability to speak, to act, to recount, to hold oneself accountable.

Ultimately, how can the training of engineers, whose purpose is the deployment of skills to innovate and invent, creating the material and human conditions for a surviving world and a desirable future,[13] be anything other than political? Here, "politics appears as the essential arena for the accomplishment of human potentials" (Ricœur, 1995, p. 37), the space where existential citizenship invites the exercise of responsibility as citizens. How can this happen? Being responsible means to impute (etymologically, to "attribute"), with the added connotation of being able to report to someone to whom I am *accountable*. Being imputable means acting as guarantor for those for whom I am responsible. For example, for parents, those dependents are their children; the head of a company or political leader is responsible for their employees or their underlings. A time dimension of responsibility is brought up by Jonas, through the responsibility principle: "taking account of the long-term consequences of the decisions made in public office and by citizens in the Technology Age'" and "elevating the concept of responsibility to the rank of a new categorical imperative—that of acting in such a way that a future humanity is still possible after our own time, in the environment of a habitable Earth" (Ricœur, 1990, p. 342). According to Ricœur's interpretation of Arendt, this exercise of responsibility involves institutional mediations in relation to each of the domains in which the capable subject is able to take part (Prouteau, 2021): the institution of language, for the speaking subject (a language marked by a history, a culture and the linguistic rules of a human community with its own traits and rules); organizations, which structure interaction (technical, monetary and fiscal, legal, pedagogical, scientific, cultural, etc.) for the acting subject; the history of human communities (associations and institutions, peoples, nations, etc.) in which each person's history is an integral part; and the respect for agreements of all sorts (law), which give structure to the exchange of given words, leading everyone to be responsible.

Freedom and human agency, political theory and freedom pedagogy, critical awareness and participation are key concepts of existential citizenship of Arendt's

[13] 'Contributing to the construction of a more desirable future for our young people' is one of the key targets of the report *Former l'ingénieur du XXIe siècle*, produced by The Shift Project (https://theshiftproject.org>uploads>2022/06)

political education, and also in Paulo Freire's work (Rüschenpöhler, 2024) which inspired the *Service-learning* through self-transformation in the world and existential self-training (Vendé & Magnin, 2023; Prouteau, 2024). This existential component is only just beginning to be taken into account in France's Grandes écoles, which are increasingly embracing service learning, especially those where social innovation and civic engagement are already core values. At the same time, they are facing up to the challenges of training teacher-researchers in these new teaching methods, which go beyond theoretical and technical learning while integrating them into citizenship education. «For apart from inquiry, apart from the praxis, individuals cannot be truly human. Knowledge emerges only through invention and re-invention, through the restless, impatient, continuing, hopeful inquiry human beings pursue in the world, with the world, and with each other." (Freire, 1970, p. 72).

In conclusion, as we bring these considerations to bear on the notion of existential citizenship, they help make explicit the multiple aspects by which a "profound existential reflection on the human condition'"(Arnsperger, 2010, p. 40) might manifest.[14] Engineering education informed by existential citizenship entails a commitment to questioning what possibilities are open to guide "a slow modification of the common spaces (the commonly shared representations) about human beings and society" gradually to legitimize "a different way of thinking and acting" (Flahault, 2005, p. 382), in order best to prepare for the future. We ourselves, in our pedagogical practices directed at students who are in a disorientating situation (Wallenhorst, 2015) or struggling to imagine a worthwhile future for the human adventure, seek to identify the emergence of action (in its arendtian sense), manifesting the political anthropology of existential citizenship. This is marked by educability—notably, political educability—as opposed to a techno-scientific anthropology of ineluctability (clearly identified in geo-scientific articles on the Anthropocene). What it demands is to create opportunities for students to participate in their own education, by offering them the right to do so and the tools to take action with respect to it. Holding such an end in view can, without hesitation, help make actors out of students, that is: subjects of *action*—something they are called to be as engineers and as students.

References

Ardoino, J. (1999). *Education et politique*. Anthropos.
Arendt, H. (1961). *Between past and future. Six exercises in political thought*. The Viking Press.
Arendt, H. (1974). *Rahel Varnhagen: The life of a Jewish Woman*. Harcourt Brace. (orig. ed. 1957).
Arendt, H. (2018). *The human condition*. University of Chicago Press. (orig. ed. 1958).

[14] In line with Masschelein, we may wonder whether "a thought and a pedagogical action, today, can still be based on the question of what we wish (or ought) to be", or whether it would be preferable to base our reasoning on other people, and what they expect of us (Masschelein, 2002, p. 47)—which lends itself to the notion of citizenship, but not that of *Bildung*.

Arnsperger, C. (2010). Changer d'existence économique : enjeux anthropologiques de la transition du capitalisme au post-capitalisme. *Revue d'éthique et de théologie morale, 258*, 23–50.

Arnsperger, C. (2011a). *L'homme économique et le sens de la vie*. Textuel.

Arnsperger, C. (2011b, October 3). Économie existentielle, critique sociale et plasticité anthropologique. In *Text accompanying the seminar "Anthropologie et Justice Sociale" at the Loyola Paris Faculties*.

Brougère, G., & Ulmann, A.-L. (Eds.). (2009). *Apprendre de la vie quotidienne*. PUF.

Dumont, L. (1991). *L'idéologie allemande*. Galimard.

Flahault, F. (2005). Vers une nouvelle pensée sociale. *Revue du MAUSS, 26*, 377–382.

Freire, P. (1970). *Pedagogy of the oppressed*. Penguin Books. (orig. ed. 1921).

Graf, W. F. (2002). Le politique dans la sphère intime, Protestantisme et culture en Allemagne au XIXème siècle. *Annales. Histoire, Sciences Sociales, 3*, 773–787.

Jonas, H. (1990). *Le principe responsabilité*. Cerf.

Jonas, H. (1998). *Pour une éthique du futur*. Payot & Rivages.

Lamarre, J.-M. (2002b). Fichte et l'éducation: devenir homme parmi les hommes. *Le Télémaque, 21*, 65–80.

Liedman, S. E. (1997). *I skuggan av framtiden – Modernitetens idéhistoria [In the shadow of the future – History of modernity]*. Albert Bonniers Förlag.

Maigné, C. (2002). J. F. Herbart: Pédagogie humaniste et critique du sujet. *Le Télémaque, 21*, 51–64.

Masschelein, J. (2002). Education et humanisme? *Le Télémaque, 21*, 37–50.

Motzkin, G. (2001). Love and *Bildung* for Hannah Arendt. In *Hannah Arendt in Jerusalem* (pp. 281–291). https://doi.org/10.1525/California/9780520220560.003.0018

Murard, N., & Tassin, E. (2006). La citoyenneté entre les frontières. *L'Homme et la société, 160-161*, 17–35.

Pineau, G. (2003). Vers une anthropo-formation en deux temps trois mouvements. *Spirale, 31*, 35–46.

Platon. (1997). *Protagoras*. Flammarion.

Prouteau, F. (2021). L'individu dans la cité: réflexion sur la citoyenneté selon Paul Ricœur. In N. Wallenhorst & E. Mutabazi (Eds.), *D'une citoyenneté empêchée à une éducation citoyenne* (pp. 175–200). Le Bord de l'eau.

Prouteau, F. (2024). *Anthropology and courage of education. Looking up the anthropocene*. Springer Nature.

Ricken, N. (2002). La philosophie de l'éducation de langue allemande. Rupture avec l'un: différence – pluralité – socialité. *Le Télémaque, 21*, 121–142.

Ricœur, P. (1990). *Soi-même comme un autre*. Seuil.

Ricœur, P. (1995). *Le Juste*. Esprit.

Rosa, H. (2013). *Accélération. Une critique sociale du temps*. La Découverte.

Rüschenpöhler, L. (2024). A review of science teaching approaches for equity focusing on race, class, and religion from the perspectives of Freire's and Arendt's theories of education. *Science Education, 108*, 1191–1221. https://doi.org/10.1002/sce.21868

Sjöström, J., Frerichs, N., Zuin, V. G., & Eilks, I. (2017). Use of the concept of *Bildung* in the international science education literature, its potential, and implications for teaching and learning. *Studies in Science Education, 53*(2), 165–192.

Tassin, E. (2004). Le cosmopolitique à l'épreuve de l'étranger. In H. Vincent (Ed.), *Citoyens du monde: Enjeux, responsabilités, concepts* (pp. 269–282). L'Harmattan.

Travadel, S., & Guarnieri, F. (2021). *Petite philosophie de l'ingénieur*. PUF.

Vendé, Y., Magnin, Th. (2023). Service-learning et transformation humaine. *Études*, 2023/12, 57–66. https://doi.org/10.3917/etu.4310.0057

Vioulac, J. (2022). Anarchéologie. *Fragments hérétiques sur la catastrophe historique*. PUF.

Vioulac, J. (2018). *Marx. Une démystification de la philosophie*. Ellipses.

Wallenhorst, N. (2015). L'étudiant face à la désorientation. In M.-H. Jacques (Ed.), *Les transitions en contexte scolaire* (pp. 281–293). PUR.

François Prouteau Professor and Deputy President-Rector for Humanities at the Catholic University of Lille (France). He holds a postgraduate degree in engineering awarded by the French Grandes Ecoles Institut Mines-Telecom Atlantique. He is a Doctor of Educational Sciences. Since October 2023, he has been the Director of the Think Tank "Meaning and Purpose at Work". Latest books: Odyssée pour une Terre habitable (Le Pommier, 2021); Anthropology of the Anthropocene. Ideas for a Courageous Education (Springer, 2024).

Nathanaël Wallenhorst Professor and Dean of the Faculty of Education at the Catholic University of the West (UCO). He is Doctor of Educational Sciences and Doktor der Philosophie (first international co-supervision PhD), and Doctor of Environmental Sciences and Doctor in Political Science (second international co-supervision PhD). He is the author of twenty books on politics, education, and anthropology in the Anthropocene. He is the leading contributor of the collections "Anthropocene - Humanities and Social sciences" (with Prof Christoph Wulf). Latest's books: A critical theory for the Anthropocene (Springer, 2023), Handbook of the Anthropocene (Eds, with Christoph Wulf, Springer, 2023).

Renaud Hétier Professor at UCO (Angers, France). A former primary school teacher and trainer, he holds a PhD in Educational Sciences from the University of Lyon 2 and a post-doctoral degree (accreditation to supervise research) from the University of Nantes. His work focuses on attention, presence and mediation in education, the anthropological and psychological reach of digital habits, and preparing for the future in times of meltdown. His notable publications (in French) include La liberté à corps perdu. Retomber sur Terre (Freedom with impetuosity. Falling back to earth), Le Pommier, 2022; L'humanité contre l'Anthropocène. Résister aux effondrements (Humanity versus the Anthropocene. Resisting the collapses), PUF, 2021; Cultiver l'attention et le care en éducation: à la source des contes merveilleux (Nurturing attention and care in education. The source of fairy tales), PURH, 2020; Créer un espace éducatif avec les contes merveilleux ou comment penser le conflit (Creating an educational space with fairy tales, or how to think about conflict), Chronique Sociale, 2017; and L'éducation, entre présence et médiation (Education, between presence and mediation), L'Harmattan, 2017.

Part II
Intersections Between *Bildung* and Efforts to Broaden Engineering Education

Chapter 7
Bildung: Conformity or Transformation?

Eddie Conlon

Abstract This chapter aims to explore how the concept of *Bildung* could be used to build an argument for a transformational approach to engineering education. Given the challenges facing humanity and the planet, education needs to reorientate towards humanizing our world. We need to oppose the neo-liberal pressure to instrumentalize education for the benefit of corporate actors. It will be argued here that the concept of *Bildung* can help with this task if it is linked to transformational models of education which aim to prepare students to help to emancipate humanity from the ravages of capitalism. Education either breathes conformity to the current system or seeks to aid those who want to change it. To change it we need to focus on a philosophy of practice which valorizes deep and active learning and links it with efforts to change extant social relations. Our objective then should be to educate activist engineers. Some barriers to doing so are also explored.

Keywords Engineering education · *Bildung* · Transformational education · Activist engineering · Social and critical realism

7.1 Introduction

A key motivation for the development of the field of engineering education research is gaining a better understanding of the change needed in the way engineering students are educated and to improve learning in a rapidly changing and crisis ridden world. I have previously argued that there are tensions between those who want change and those who are committed to the status quo and have a preference for the

---

E. Conlon (✉)
School of Transport and Civil Engineering, TU Dublin, Dublin, Ireland
e-mail: edward.conlon@TUDublin.ie

A. Buch, S. H. Christensen (eds.), *Bildung for Engineering Education and Practice*, Philosophy of Engineering and Technology 49,
https://doi.org/10.1007/978-3-031-86581-7_7

engineering "core" and knowledge which has a mathematical content at the expense of almost everything else in the curriculum (Ramnath et al., 2023, see also Murphy et al., 2022). But those who want change are often unclear about what it is, that needs to change and why? They are often unclear about what we are trying to do when we educate engineers.This tends to pull attempts to change engineering education "into opposing, even contradictory, directions" (Jamison et al., 2014, p. 254).

Jamison et al. (2014) have told us that historically there are contesting models of engineering education in which dominant approaches, which emphasizes the technical and entrepreneurial dimensions of engineering education, have marginalized more integrated or social approaches focused on the public good and role of engineers as change agents. The result is that students "have not been given the opportunity to understand the broader social and cultural aspects of the challenges facing engineering" (Jamison, 2013, p. 21). In my own work I have argued that the "social dimension" of engineering education has incorporated two contrasting aims for engineering education: one focused on developing students as employable graduates with developed social and communications competencies and the other embracing a wider role for graduate engineers as ethical change agents capable of fighting for a more sustainable world (Conlon, 2008). Others have identified a superficial approach with efforts to integrate the social into engineering education being "driven by surface-level conceptualizations and a lack of deeper understanding". (Xavier et al., 2022). Efforts to expand engineering education to incorporate ethical issues tend to treat the social context of engineering in a superficial way failing to give students the knowledge they need to both understand and change the world and engage in what Bhaskar, the founder of Critical Realism, calls *depth struggle* (See Bhasker, 2014; Conlon, 2022). This involves a commitment to *explanatory critique*: an explicit commitment to understanding the deeper causes of things to transform them:

> Hence the production of explanations of social institutions is not only, as a general rule, a precondition of criticizing and changing them; sometimes, it *is* criticizing them and beginning the work of their subversion (Collier, 1994, p. 172).

Collier points to Marx's exposure of the exploitative nature of the wage system as a classic example: "To expose is to criticize…and to spread the word is to stir up dissent from capitalism, which of course is just what Marx intended" (p. 173).[1]

This points to the importance of linking theory and practice and a *philosophy of praxis*, derived from the work of Marx, and which has influenced radical educators such as Freire (1972). It calls for "a dialectical synthesis of theory and practice orientated towards transforming extant social relations" (Scatamburlo-D'Annibale et al., 2018, p. 550).

[1]Allman et al. (2005) illustrate how much analysis of social inequality is not committed to such explanatory critique:"in general, the explanation of social inequality has not been undertaken as a form of critique of existing social relations. In this sense, it has not been embarked upon seriously. At stake is the uncovering of the forces that generate social inequality. This type of analysis is not content just to describe and tabloidise (in postmodernist or liberal Left mode) the empirical manifestations of these forces… Such a critique must go to the roots of capitalist society" (p. 139).

This chapter aims to explore how the concept of *Bildung* can be used to build an argument for a transformational approach to engineering education to help resist pressures towards instrumentalization and provide a students with the kind of knowledge they need to prepare them to take part in the struggle for social justice. It will argue for the need to link the concept of *Bildung* with radical approaches to education which seek to transform educational practices to change our world. It will also explore barriers to the development of such an approach. It can be stated at this point that the commitment to inserting graduates in to the runaway of globalized capitalism (Ramnath et al., 2023) in higher education presents a real barrier to the development of a genuinely transformational approach. I agree with Schaull (1972), as argued in his Foreword to Freire's *Pedagogy of the Oppressed*, that educators face a simple choice:

> There is no such thing as a neutral educational process. Education either functions as an instrument which is used to facilitate the integration of the younger generation into the logic of the present system and bring about conformity to it, *or* it becomes 'the practice of freedom', the means by which men and women deal critically and creatively with reality and discover how to participate in the transformation of their world (p. 14).

The question we need to ask is: what do we need to do to ensure that education becomes the practice of freedom?

7.2 Into the Deep[2]

A key issue revolves around how we understand the purpose of education. The lack of depth identified above can be traced to the instrumentalization of knowledge within education. Wheelahan (2010), and others working within a Social Realist framework (Moore, 2013; Priestly, 2011; Young, 2008), have decried the death of knowledge in the curriculum and argued that while new approaches to education seem to challenge traditionalists both approaches share an instrumental, and conservative approach to knowledge. Traditionalists in education have often defended access to knowledge as a mechanism of social stratification and selection. They value less the intrinsic role of knowledge in society and are more concerned with

> enculturation into traditional values and norms, based on a relationship of deference to traditional bodies of knowledge taught in traditional authoritarian ways that require submission to become the 'kind of person it is supposed to make you' (Wheelahan, 2010, p. 7).

In engineering, the foundational knowledge, seen to be at the core of engineering, which has made engineers what they are, is seen to be under threat from constructivist approaches committed to context dependent learning and to real world problem

[2] The focus on depth is inspired by Marx's observation that there would be no need for science if the outward appearances and essences of things coincided. Small (2014) argues that the essence of Marx-the-educator was to challenge us "look beneath the surface of (things) and find out what is really happening" (p. 2, see also Conlon, 2022).

solving entailing the use of teamwork and other skills emphasized by new approaches to engineering education. New approaches are influenced by both constructivism and instrumentalism as both share a common focus on context: "both emphasize the contextual, situated and problem-orientated nature of knowledge creation and learning (and) sacrifice the complexity and depth of knowledge in curriculum in favor of 'authentic' learning in the workplace." (Wheelahan, 2010, p. 5). So, while these new approaches may seem to represent a break with the traditionalists, Wheelahan (2010, p. 7) argues that they all share an instrumental attitude to knowledge and do not see it as "a causally important objective in its own right because of the access it provides to the nature of the world and to society's conversation". She defends the centrality of abstract theoretical knowledge to educational endeavors. It's the means that society "uses to think the 'not-yet-thought and unthinkable' and to imagine alternative futures… and make connections between objects and events that are not obviously related". She links a commitment to deeper understanding of the world with efforts to change it.

In a somewhat similar manner those committed to transformative learning (e.g. Sterling, 2011) argue for an education that "take us into the depth of things" and is critical of reforms in education which focus only on learning methods with much less attention on the *purpose of learning*:[3]

> So, whilst learning skills, or 'learning to learn' may be important…this emphasis does not necessarily address the critical dimensions of context or reflexivity. Beyond 'learning to learn' is 'learning about learning', which raises questions about validity and ethical defensibility of our theory and practice (p. 19).

Transformative leaning requires a qualitative shift in our assumptions, habits or thoughts, a changing understanding of power relations and a sense of possibilities for social justice and change. According to Sterling this requires attention to 'levels of knowing' and how deeper perceptions and conceptions influence more immediate ideas and actions. Transformative learning thus requires a dramatic shift in consciousness and thus the concept has similarities with Paulo Freire's (1972) concept of conscientization which has significant influence in emancipatory education circles. The overall goal is social change and the linking of transformations in the way things are understood with transformative action.

7.3 What Needs to Change?

What follows from this is that we have a challenge. How do we develop and implement a non-instrumental approach to education, a commitment to knowledge as an end in itself, with the aim of giving students the powerful and critical knowledge

[3] Sterling (2004) has analysed higher education using an iceberg metaphor and argues that "the deeper levels of paradigm and purpose guiding policy and practice…tend to be hidden from view and… most debate" (64).

they need to play an active role in addressing the substantial challenges facing engineering and the planet engineers share with others (Ramnath et al., 2023).

The challenges are substantial and are fundamentally linked to the nature of the economic system under which we live. Valerie Scatamburlo-D'Annibale et al. (2018) argue that the fundamentally exploitative basis of capitalism is rarely addressed in dominant educational discourses which ignore the historical specificity of the capitalist mode of production and the manner in which education leads students to internalize "the demands of the methods of production as self-evident laws of nature" (p. 553).

Even those who would call for the "disruption of engineering education" would seem to accept this. So, they pose the challenge for education in engineering in the following terms:

> We need our graduates to be able to continuously think outside the box, upskill in emerging technologies, processes, and systems, all while meeting the day-to-day demands of their jobs. We strive to produce engineers of the future for workplaces of the future. So how do we continue to innovate engineering education to produce these future-proofed graduates? (Lindsay et al., 2023, p. 115)

This would seem to suggest engineering education should continue to integrate graduates into the logic of the current system. A system based on the exploitation of labor, constant growth in pursuit of profit and universal commodification which tends to reduce all activity to the law of value with deleterious consequences for the planet and the people on it. Some of the worst of these can be listed as:

1. The record for the hottest year was "smashed" in 2023 by a huge margin. The planet was 1.48 degrees hotter compared with the period before the mass burning of fossil fuels ignited the climate crisis. The figure is very close to the 1.5 C temperature target set by countries in Paris in 2015 leading to a requirement to rapidly cut fossil fuel use to preserve the livable climate that we all depend on (Carrington, 2024).
2. Despite this, and because of the need to realize the value of assets, fossil fuel investment spending is now more than double the levels needed to reach net zero emissions by 2050. Between 2020 and 2023 investment in fossil fuels increased.[4]
3. Seven of the eight earth-system boundaries have now been pushed beyond their safe and just limits into risk zones that indicate the threat to planetary and human health (Mooney, 2023).
4. While there is now a significant focus on the Sustainable Development and the Sustainable Development Goals in engineering education (see Beagon et al., 2023; Nicolaou et al., 2018)[5] we were told in 2023 that 8 years after their adoption only 15% of the targets are on track. Millions of people have fallen into

[4] https://www.iea.org/reports/world-energy-investment-2023/overview-and-key-findings, accessed 10 March 2024.

[5] The Eleventh International Conference on Engineering Education for Sustainable Development was held in 2023. see https://eesd2023.org/ (accessed 10 March 2024).The accreditation criteria for engineering programs in Ireland states that" Students should be exposed to, for example, the

poverty, hunger and malnutrition are becoming more prevalent, humanitarian needs are rising, and the impacts of climate change are more pronounced (Wintour, 2023).

5. While the UN declares that the world is far off track to meet the goal of ending hunger by 2030 global inequality has increased. According to Oxfam, in 2023 the richest 1% grabbed nearly two-thirds of all new wealth worth $42 trillion created since 2020. During the past decade "the richest 1% captured around half of all new wealth" (Oxfam, 2023). Given the historical entanglement of capitalism with sexism and racism it is perhaps unsurprising that recent data,[6] for 140 countries, shows that that 70% of countries saw racial discrimination worsen between 2021 and 2022. Since 2015, discrimination has increased in three quarters of countries included. The global gender gap score[7] improved by only 0.3% points between 2022 and 2023. When considering the 102 countries covered continuously from 2006 to 2023, the gap was 68.6% closed in 2023. At the current rate of progress, it will take 131 years to reach full equality for women.
6. Finally, the *share* of labor income in GDP globally has fallen since the mid-1970s with gains for capital throughout the globe. As Kim Moody puts it: "In short, the working class has lost out everywhere". Worker organization has weakened in many places, while "advanced methods for the surveillance, measurement, quantification, and standardization of work that ultimately impact workers everywhere" have boosted return to capital and made the experience of work more stressful for workers (Moody, 2021).

These challenges are significant, and engineers are implicated in many of them. Further, capitalism is subject to recurrent crises that have begun to take place more regularly, and are played out in a highly integrated global system within which their effects are more… devastating than ever before" (Allman, 2001, p. 162). As a result Allman argues, that "there is no longer room in critical education for hesitation or prevarication". We need to renew critical education and engage with the struggle for social justice, or "we capitulate to capitalism and the continuing dehumanization of millions of human beings" (ibid. 2001, p. 162).

It is not at all clear that engineering education, or education in general, is preparing students to address these challenges. Broadly the issues listed above raise questions related to sustainable development in that they focus on issues of inter- and intragenerational equity and the carrying capacity of the planet. Yet there is evidence that efforts to integrate sustainability into education generally or engineering education specifically have failed and failed because the education being offered to students is captive to the very system that produces unsustainability in the first place.

UN Sustainable Development Goals, the 'blueprint to achieve a better and more sustainable future for all', consisting of 17 goals to be achieved by 2030." (Engineers Ireland, 2021, p. 17).

[6] https://worldjusticeproject.org/news/discrimination-getting-worse-globally (accessed 17–03-24)

[7] https://www.weforum.org/publications/global-gender-gap-report-2023/digest/ (accessed 17–03-24)

So Huckle and Wals judge, that the Decade of Education for Sustainable Development to have "failed to acknowledge or challenge neo-liberalism as a hegemonic force blocking transitions towards genuine sustainability" (Huckle & Wals, 2015, p. 491). One leading advocate of Engineering Education for Sustainable Development (EESD) has claimed that:

> Others have concluded that EESD is not happening at the pace it should…The commodification of higher education has also been highlighted as a barrier and impediment to a clear institutional commitment to the Barcelona Principles[8] with university rankings (and the metrics which underlie these) becoming increasingly more important for measuring universities global competitiveness (Fenner, 2023).

While another relates the lack of progress to the focus in higher education on the demands of the labor market: "The result of these developments was that engineering students were, during their studies, increasingly focusing on their professional qualification…for a career as an (engineering) specialist." (Mulder, 2019).

Outside of engineering education others have been harsher claiming that, as they are currently constituted, "universities are complicit in reproducing unsustainability and inequality" (McGeown & Barry, 2023, p. 2) as they operate in the interests of neo-liberal capitalism to which they are captive. While neo-liberalism certainly represents a most aggressive form of capitalism, the captivity of education to the economic system is not a new problem.

In general terms capitalism faces the problem of the reproduction of each new generation of workers with the skills but also the willingness to utilize their abilities to expend themselves as a value-creating force, Education plays a central role in this and in producing well motivated and employable graduates with the right attitudes to work (Allman et al., 2005).Critics of capitalism have pointed out how the more knowledge became a productive power, the more its creation and acquisition became an organic part of commodity production and education was determined by the demands of the wage labor system (Allman, 2001; Noble, 1977; Pavlidis, 2013). This diminishes "any deeper meaning education could have for human life". (Pavlidis, 2013, pp. 299–300). So, it seems in order to change education, we need to understand, that its dominant purpose and content is shaped by the material and social conditions of human life.

But in doing so we need to adopt a non-determinist approach. I have argued elsewhere (Conlon, 2019) that the structures of capitalism do not determine exactly what people do but rather shape behavior by setting boundaries and limits. This leaves open the possibility that humans as agents develop consciousness of the imposition of structures on their lives and can take action to change them. Further, the contradictory role of the education system in reproducing inequality, while at the same time legitimating ideologies of equality, allows room for resistance to this reproduction (Au, 2014).[9]

---

[8] This refers to the Barcelona Declaration on EESD.

[9] Au (2014) is an excellent discussion of a non-determinist Marxist approach to education. Banfield (2015) says that "if human agents were erased from history, then a Marxian conception of educa-

7.4 Can *Bildung* Help?

The question arises if a focus on *Bildung* can help in providing our students with the kind of education they need to humanize the world. *Bildung* is one of the philosophical traditions, which also includes *paideia* and *liberal education* that have argued for treating education as an end in itself and the promotion of its human dimensions (Pavlidis, 2013). A commitment to education in its own right can reveal to us "dimensions of education, that allude to another approach to the meaning of life, which is radically different from the currently prevalent perception (ibid., p. 300). Thus, a focus on the ideal of education for its own sake can help with the development of alternative conceptions of education. Furthermore, in the context of policy discourses focused on standards, competencies, and benchmarks and with education systems becoming increasingly "vessels for economic growth" (Restad, 2019, p. 402) the concept of *Bildung* has experienced a revival as a counter-concept (Sjöström et al., 2017) as "it is objectively in opposition of the contemporary commercialisation of educational good and activities" (Pavlidis, 2013, p. 300). So *Bildung* may help but there are some issues to be addressed.

First is the meaning of *Bildung* and whether it supports efforts to develop a transformational approach to education. While in a general sense it can be seen to focus on self-realisation or cultivation; "the development of all universal human powers and at the same time as the cultivation of all unique characteristics of an individuality" (Pavlidis, 2013, p. 301) some have argued that the concept is "vague, ambiguous and difficult to put into practice" (Bohlin, 2008, p. 2). It is also a contested concept which may support traditional approaches to education.

Drawing on historical research, Moore (2004, Chap. 3) points out that the concept was associated with efforts to cultivate a middle class who would occupy the German state apparatus and preserve order and stability. Thus, the *ideology of cultivation* was an agent of social stratification and differentiation. Indeed, one finds in the history of engineering education a focus on cultivation through efforts to enhance the status of engineers by exposing them to courses in the liberal arts and social sciences to provide them with the well-rounded education worthy of "gentlemen" (sic) (Seely, 2005). But debates about broadening the education of engineers were not devoid of concerns about what David Noble calls "the business of living". The incorporation of "cultural studies" into the curriculum (in the USA at the end the 19th and start of the twentieth century) came to be seen by engineering educators to have practical value as well as "mere status value" enabling engineers to "meet men (sic) as well as matter" and give engineers, who increasingly played managerial roles in industry, "greater power over men (sic)" (Noble, 1977, pp. 31–32). What can be taken from this is the importance of having a clear rationale for efforts to broaden the education of engineers and whether they contribute to conformity or transformation.

---

tion, or any conception of education, would be impossible" (p. 46).

A related problem is the danger that the ideal of education, if overly focused on individual cultivation, could be detached from the material world, people's productive activity within it and the exploitation and alienation emanating from class society which generates the conditions for the subjection of education to utilitarian purposes in the first place. An education pursued for the sake of individual cultivation risks. Pavlidis (2013) argues, being prone to "an uncritical acceptance of the currently dominant forms of knowledge and culture". Only when individual life goals encounter universal social needs does the individual interest in education transcend the horizons of individual satisfaction to focus on "the continuous effort for a deeper understanding of the human condition, and of the needs, achievements and potential of humanity" (Pavlidis, 2013, p. 305). It is perhaps not surprising that the contemporary use of *Bildung* has an emphasis on autonomous self-formation *and* emancipation (Sjöström et al., 2017). But even then, we need some clarity of what we need to be emancipated from. Sjöström et al. (2017) want to characterise modern society, following Beck (1992), as a "latemodern risk society" (p. 180). However, such a characterisation fails to appreciate the central role of capital in generating risk and the differential distribution of risk based on traditional cleavages of class (Pearce & Tombs, 1998).

7.4.1 What Does It Mean?

According to Sjöström et al. (2017) *Bildung* has a variety of definitions, at least five. Bohlin (2008) cites work which identifies twelve main meanings as used in contemporary Swedish educational debates. He notes though, that the long list of meanings reveals both the ambiguity of concept but also the "importance of many of the ideas associated with it for any serious debate on the general goals of education" (p. 2). He claims that many of the classical thinkers in the tradition considered the "development of moral autonomy and responsibility and a readiness for morally motivated action to be the most important aspects of the *Bildung* process" (2008, p. 3) thus suggesting some role for it in developing a transformational approach.

So, what does it mean? Sjöström et al. (2017) suggest some basics: the concept hinges on the idea that knowledge fundamentally changes human beings; by learning certain knowledge and skills their agency is enabled by allowing the person to become an active subject in the world rather than being externally determined. It also includes, nowadays, a commitment to critical knowledge, value consciousness, ethics and responsibility with, under the influence of *Critical Theory,* a commitment to emancipation defined as eliminating oppression and creating the conditions for effective agency. As such it should not be understood individualistically but it is about the individual being embedded in the world. Sjöström et al. (2017, p. 178) opt for a version which they call Critical-reflexive *Bildung* involving a commitment to critical consciousness, critical literacy and critical agency which involves action to promote social justice. This amounts to what they call *critical praxis* or reflection in practice. They emphasize how it is not possible to fight oppression without expanded understandings and perspectives.

This connects the concept of *Bildung* to contemporary emphases, for example in sustainability education, on transformative learning (Sterling, 2004, 2011). Bentham (2013) argues that *Bildung,* much like action, competence, and the ideal of sustainability education, values the development of the reflective individual, who has the power "to question assumptions, 'facts', agendas and opinions about current living conditions and activities" (p. 36). Bohlin (2008) links it explicitly with transformational learning methods arguing that these provide effective tools to implement moral *Bildung* in educational practice, while Torres and Yan (2022) connect it to the work of Freire, and its role "in nurturing democratic citizenship in an era of new authoritarianism" (2023, p. 660).

Thus, in contemporary debates it is worth noting the linking of *Bildung* with radical and transformational approaches to education which aim to counter the subjugation of education to the needs of the economy and address the unsustainability of modern capitalism. The aim of such an education should be, in Freire's (1972) terms, "the humanization of the world". Thus, a commitment to *Bildung,* understood as providing students with the depth knowledge they need to become critical and reflective actors for transformational change, may act as a focus for change efforts in engineering education. It may also aid those resisting an overly instrumental approach focused only on producing graduates with the skills and attributes needed for a changing workplace.

Drawing on some work on transformational education (Allman, 2001; Bohlin, 2008; Freire, 1972; Karwat et al., 2015; Scatamburlo-D'Annibale et al., 2018; Sterling, 2011) we can identify what this means in practice:

1. A focus on "paradigmatic reconstruction": a shift in our ways of knowing and understanding and "seeing our worldview rather than seeing *with* our worldview" (Sterling, 2011 p. 23). We need to challenge the normalization of the logic of commodification and the pursuit of profit as the basis for organizing society (Scatamburlo-D'Annibale et al., 2018). In engineering it requires us to challenge the idea that we can "technologize" (Karwat et al., 2015) our way out of the problems we face and acknowledge that the current economic system is sustained by and dependent on continuous technological development. As educators we need to subvert the "smooth flow of labour-power production by inserting principles antagonistic to the...domination of capital" (Allman et al., 2005, p. 157). These include social justice, equality and solidarity with progressive social struggles.
2. Combined with this we need to reject dominant notions of apoliticism and ahistoricism in engineering education and practice (Karwat et al., 2015; Murphy et al., 2022). This requires providing students with a deeper understanding of the causes of unsustainability (as being linked to the operation of the economic system rather than to the behaviour of individuals) with a view to helping them envision real alternatives. Joel Spring has summarised Freire's understanding of humanisation as:

> a process by which each person becomes conscious of the social forces working upon him or her, reflects upon those forces, and becomes capable of transforming the world. To be human is to be an actor who makes choices and seeks to guide one's own destiny. The opposite of a humanized world is one without self-awareness, without a consciousness of the historical forces determining existence. Without this consciousness people are unable to become actors in the stream of history and are simply acted upon by history (Spring, 1975, pp. 62–63).

3. A focus on empowering students to learn and transform them from being objects of educational processes and receptacles of knowledge via what Freire calls the "banking method" of education. Passive students do not become social activists. We can support efforts to develop active learning in engineering education but with the purpose of what Bohlin (2008) call*s moral self-cultivation.* A process that involves:

> realising alternative to one habitual way of thinking, feeling, and acting, thereby (ii) becoming able to identify the presuppositions or assumptions underlying those ways of thinking, etc., which in turn leads to (iii) critical assessments of these presuppositions-and…(iv) critical assessment of the alternatives (pp. 8–9).

These alternatives must include proposals for changing extant social relationships and structures to enable the moral agency of engineers (Conlon, 2022).

A commitment to active learning does not mean we turn teachers into passive facilitators. As McClaren (2001) points out, Allman has argued that it is prescriptive to "ask students to read the world critically in order to transform it". Furthermore, "shouldn't educators use their authority that comes from their own critical reading of the world", he says:

> Isn't the most facilitative, non-prescriptive and non-directive form of progressive teaching doubly prescriptive in the sense that it is a prescription for non-prescription as well as for political domestication and adopting successfully to the social universe of capital and the law of value…Freirean educators direct and prescribe but in a way redolent of humility and a spirt of mutuality (p. x).

While encouraging active learning and dialogue with students, we need to understand that education is never just a process. We always need to think about what it is we want students to learn and why (Young, 2008). So, we also need to expose students to what social and critical realists call powerful and critical knowledge that reveals the deeper structure at work in the world (see Conlon, 2022; Huckle, 2017; Young, 2008).

4. All the above should lead our students to engaging in struggle to humanise the world and become "activist engineers" (Karwat et al., 2015). As Scatamburlo-D'Annibale et al. (2018) have argued contrary to liberal reformists, who tend to concentrate on "conscious raising", Freire had little use for forms of social activism that leaves the fundamental structures of society untouched. For him humans do not get beyond the situation they find themselves in only because of their intentions or consciousness. Verbal denunciation of the system is not enough. As

I have argued elsewhere (Conlon, 2019, 2022), we need to convey to students how, by working with others, such as social movements and trade unions, engineers can be crucial participants in larger social and political movements to challenge accepted directions and choices in technology.

7.5 A Useful Education?

A key problem in moving towards a transformational model of engineering education is that, the subordination of education to utilitarian purposes, as suggested above, "is not an issue of individual desires and choices (but) is an inevitable, prevalent rule of capitalist society, in which man (sic), as a waged worker, can exist only if he can find a buyer of his labour power" (Pavlidis, 2013, p. 304).

The subjugation of science, and engineering, to productive purposes makes this problem more acute for those who must educate engineers and scientists. Indeed, it might be argued, especially for those involved in the education of the professions, that an education that does not lead to the preparation of graduates for the workplace is not of much use to them. Barnett (2009) identifies a move in higher education to a *dogma of skills* and *the performative student* in line with the increasing neoliberal functionalism and economisation of higher education. Given this emphasis and the financial pressures on students Bohlin argues that:

> even among those who would prefer if possible to devote at least part of their higher education to *bildende* topics of study, many will choose not to do so. In other words, the freedom which is necessary of the higher stages of moral self-development can easily lead to a situation where little of no moral self-development takes place (2008, p. 5).

But an education that only focuses on *doing* may not be of much use to graduates or society. Students need access to theoretical knowledge to realise their rights to participate in conversations in society and in their fields of practice about what they have done in the past and what they should do differently in the future and why. This takes on added significance given the increasingly complex nature of work and technology (Wheelahan, 2015). Pavlidis argues that education needs to deal with a broad spectrum of scientific-technological issues so that graduates can deal with radical and scientific developments flexibly and independently and "consider the moral-social dimensions of science, those that relate to the impact of scientific and technological applications on the existence and evolution of humanity" (Pavlidis, 2013, p. 307). As Science and Technology Studies (STS) scholarship tells us, as engineers create or develop technology they are "intimately involved in maintaining key social patterns and inventing new ones as well" (Winner, 1990, p. 58). Engineers are not just cobbling artefacts together but rather they are building the framework and social structure for society in which humans live (Johnson & Wetmore, 2021).

So being a professional has both technical, social, and moral aspects (Fellenz, 2016). Given their autonomy, the impact their work can have on others and their role in contributing to the general development of professional approaches and standards professionals need more than skills. They must also be able to evolve to reflect on changing technological, social, political, ethical, and economic circumstances. Fellenz argues, that because professionals need to develop and engage in self-formation, and do so with recognition of the nature and role of the social, political, and cultural context in which they are formed, *Bildung* is "a particularly useful lens with which to consider professional education and development" (2016, p. 273).

7.6 Who Calls the Shots?

A related obstacle is that educators are not always in control of what happens in the curriculum. There are powerful actors outside the academy who shape what happens in the engineering classroom and these can often reinforce (or resist) pressures towards the commodification of education and a wholesale focus on preparing students for the labour market.

Slaughter (2002) has argued that curricular change cannot be understood by just focusing on the activities of scholars and curriculum researchers or the response of institutions to changing market forces and demographic change. A complete account of curriculum formation must take account of the activities of professional class organizations, social movements, and external resource providers. Further, Slaughter argues that strong programs have strong professional associations. They "create and maintain many of the links that connect the academy, the economy and the state" (2002, p. 283).

In Slaughter's view, direct ties between curricula and external organisations raise "the possibility of academic knowledge as the servant of power, ministering to special interests and at odds with the public good "(2002, p. 273) but also the possibility that progressive change can come from struggle around the content of education.

Slaughter usefully suggests that the curriculum is shaped by a network of actors and agencies relevant to each discipline: knowledge is contested and constructed within a context of competing and cooperating interests. What is useful about this is that it forces us to think beyond the academy and its staff to consider a range of actors who may influence the curriculum. They include professional bodies, learned associations, employer and worker organizations, state and corporate actors and social movements including student organizations.

Considering Slaughters work, it is worthwhile asking what the key actors have to say about the ideal of education and the extent to which they embrace a commitment to an education as an end in itself, and as a tool of transformational change. While a complete account of what various actors have to say on this issue should be the subject of systematic research, a cursory look at what some key actors have to say about the purpose of education suggests that the terrain in Irish higher education is somewhat unfavourable to the approach argued for here. I give three examples.

Firstly, after the election in 2020 a decision was made to establish a Department of Further and Higher Education Research, Innovation and Science (DFHERIS). The new Department, like others, is required to produce a statement of strategy every 2 years. Its statement for 2023–25 (DFHERIS, 2023) tells us much about how the Department considers the purpose of further and higher education. It is a rather grim document in considering the perspective argued for here.

It sees the role of DFHERIS as contributing to creating an" innovative, international Island with inclusion at the heart of our agenda". It says:" our vision (is) to become *an engine for economic growth,* as well as an instrument to enhance social cohesion." In light of this it says that at:" the heart of everything we do will be access and inclusion and ensuring there are no barriers to education, or employment" (p. 4). Continuing with this theme of marrying a focus on economic growth with inclusion, it sets out a strategy to enabling" everyone across society to meet their potential and develop the talents, skills and knowledge required to succeed *in the Irish economy* of today and meet the needs of our stakeholders, employers and industry." (p. 9).

In terms of the significant challenges facing us, it says, learners will be enabled to respond to the" evolving economic, social, environmental, and technological environment, in particular in rapidly growing areas such as transversal skills, climate and green skills, digital and artificial intelligence." (p. 18).

Its notable that in a 44 page document concerned with the vision and strategy for higher education, knowledge is mentioned hardly at all (6 times, two of which are to repeat previous mentions) and there is no indication of what kind of knowledge students will get in the system. In comparison, skills are mentioned 40 times as the overwhelming emphasis is on inserting graduates into the Irish economy of today. Where significant challenges are mentioned at all, it is said that students should be given the skills to respond to the challenges without any sense of the students intervening to transform the conditions that lead to these challenges in the first place.

While there is some reference to inclusion, this is aimed at integrating those on the margins into the state project of "growth and prosperity". There is no reference to the knowledge these students need to fully participate in society's conversations. Inclusion without knowledge is likely to make education a key way," in which socially inequality is mediated and reproduced" (Wheelahan, 2015, p. 751). It remains a mechanism for social stratification and the generation of conformity.

Secondly, employers play a significant role in Irish engineering education. They participate in the accreditation system operated by Engineers Ireland, the professional body for engineers in Ireland, with employer representatives sitting on accreditation panels (see Conlon, 2013). In some institutions employers act as external examiners. So, what employers have to say about the purpose of engineering education matters.

The main body representing employers in Ireland is the Irish Business and Employers Confederation (IBEC). It produced a report on improving graduate employability skills and amounts to a significant statement on the purpose of education (IBEC, 2018). It starts by arguing, that" business is not the only, or indeed the primary, stakeholder in education". However, IBEC says that:

> contained within the strategic plans, missions and visions of most of our universities, colleges and further education institutions is a focus on or commitment to supporting learners to achieve their full potential…and here lies an important area of overlap and shared concern with business…There is a responsibility and an increasing student demand to provide opportunities to help develop their employability skills and an understanding of the knowledge, skills and attributes that they can offer prospective employers. (p. 9)

IBEC claim that" the very intellectual attributes that are associated with a liberal education, for example, are the same as those, that are essential for ideas, innovation and intrapreneurship in the workplace" and that" Education should attract, excite and develop graduate talent for the future." (p. 9).

IBEC goes on to call for a wholesale integration of employability across the curriculum by providing "dedicated employability supports for all academic staff", including sending them on placements to industry, reviewing" teaching program to consider the opportunities for employability in the curriculum", increasing "the application of relevant and innovative assessment method" and making students" familiar with the language and terminology of employability" (pp. 14–18). Students should be encouraged to have" an entrepreneurial mindset" (p. 18).

Finally, it is claimed that" employability skills and attributes are the high level generic attributes that are necessary to allow Irish graduates to contribute to and thrive in a complex and uncertain future." Therefore," students, teachers, education institutions and employers need to share the responsibility of preparing future graduates through the development and communication of employability skills" (p. 26).

This is in fact a wholesale call for a reorientation of higher education. While it is acknowledged that there are other aims for higher education, and that there may be crossover between employability skills and learning skills more generally, the problem is that the purpose of education is framed around employers' needs and the continuation of growth and innovations with an overwhelming focus on skills over knowledge. As I have argued elsewhere (Conlon, 2008), the employability agenda is about getting graduates to adapt to the new flexible workplace. The skills identified as core, produce the type of knowledge and understanding that is required to maintain dominant cultural and political arrangements. A focus on employability skills will not give engineers the capacities required to reflect critically on the structure of work and the way in which the rewards of productive activity are distributed.

Finally, given the position of these two important actors it is necessary to ask if there are countervailing forces pushing for a more transformational approach. Given its central role in regulating engineering education, and the importance attached by engineering educators to it (Murphy et al., 2022; Nicolaou et al., 2018), one could look to Engineers Ireland (EI), the body responsible for the accreditation of engineering programs in Ireland. Accreditation criteria (Engineers Ireland, 2021) include a set of program outcomes against which programs are examined ranging from engineering science and mathematics, problem-solving, and design, to what some consider "softer" skills related to ethics, teamwork, and communication. In 2021, an outcome related to engineering management was added. The ethics outcome requires that students have "an understanding and appreciation of the environmental, social and economic impacts of their judgements and promote the principles and practices of sustainable development" while "understanding of the importance of the engineer's role in society and the… importance of equality, diversity and inclusion." This provides a possible basis for the broadening of the curriculum in the manner argued for earlier but, based on the research evidence,we must have doubts that the pressure exercised by accreditation bodies can translate into deeper curricular change (Martin et al., 2021 see also Ramnath et al., 2023 for a summary of the research findings).

Most harmful, the most comprehensive study we have of ethics education, which includes a substantial interrogation of the accreditation process, claims that ethics is given lower weight in the curriculum and that this is considered normal. It appears that" ethics does not need the same emphasis in the engineering curriculum compared to technical oriented outcomes" (Martin, 2020, p. 247). The study supports the view that the existence of an accreditation criterion dedicated to ethics does not necessarily lead to a curriculum which addresses the social and political dimensions of engineering practice in a broad manner, or that accreditation processes adequately interrogate the extent to which programs address this outcome. So, we face a contradiction of an aspiration to broaden engineering education being subverted by the culture of engineering which valorizes the technically proficient and the entrepreneurial. This is reinforced by the addition of the Engineering Management program outcome. This would seem to push engineers towards conformity with current economic arrangements rather than challenge them.

7.7 Conclusion

Given the challenges facing humanity and the planet, education needs to reorientate towards humanizing our world. We need to oppose the neo-liberal pressure to instrumentalize education for the benefit of corporate actors. I have argued here that the concept of *Bildung* can help with this task if it is linked to transformational models of education which aim to prepare students to help to emancipate humanity from the ravages of capitalism. Education either breathes conformity to the current system or seeks to aid those who want to change it. To change it we need to focus on a philosophy of practice which valorizes deep and active learning and links it with efforts to change extant social relations.

Given the forces united against such a change in education we need to identify those forces who would work for it. This requires developing, in a systematic way, the work started here but which examines the vision for education of a wider group of actors. In doing so, we need, in the spirit of the philosophy of practice, to link this work with action to change extant social relations within higher education. If we really want to disrupt engineering education, foster the practice of freedom, and generate dissent, our purpose, aided by a transformational vision of engineering education, should be, as Karwat and others put it, to create activist engineers:

> It is the role of the activist engineer to create a new paradigm of engineering in which the engineer is equipped with not only technical tools and knowhow, but also with the requisite socioecological perspectives, knowhow and ethics that allow for activist engineering. If engineers have been essential in building and maintaining the current sociotechnical order, they must be the ones engaged and empowered to forge a new order (2015, p. 232).

References

Allman, P. (2001). *Critical education against global capitalism.* Bergin and Garvey.

Allman, P., McLaren, P., & Rikowski, G. (2005). After the box people: The labor-capital relation as class constitution and its consequences for Marxist educational theory and human resistance. In P. McLaren (Ed.), *Capitalists and conquerors: A critical pedagogy against empire*. Rowman & Littlefield.

Au, W. (2014). Against economic determinism: Revisiting the roots of neo-Marxism in critical educational theory. *Journal for Critical Education Policy Studies, 4*(2), 11–35.

Banfield, G. (2015). Reclaiming history: Marx, education and class struggle. In G. Grollios, A. Liambas, & P. Pavlidis (Eds.), *Proccedings of the IV international conference on critical education "Critical education in the era of crisis"* (pp. 38–63) http://www.eled.auth.gr/, Accessed 16-13-24

Barnett, R. (2009). Knowing and becoming in the higher education curriculum. *Studies in Higher Education, 34*(4), 429–440.

Beagon, U., Kövesi, K., Tabas, B., Nørgaard, B., Lehtinen, R., Bowe, B., Gillet, C., & Monrad Spliid, C. (2023). Preparing engineering students for the challenges of the SDGs: What competences are required? *European Journal of Engineering Education, 48*(1), 1–23.

Beck, U. (1992). *Risk society: Towards a new modernity.* Sage

Bentham, H. (2013). Clearing the path that has been laid: A conceptualisation of education for sustainable development. *Journal of Teacher Education for Sustainability., 15*(2), 25–41.

Bhasker, R. (2014). Foreword. In P. K. Edwards, J. O'Mahoney, & S. Vincent (Eds.), *Studying organizations using critical realism: A practical guide*. Oxford University Press.

Bohlin, H. (2008). *Bildung* and moral self-cultivation in higher education: What does it mean and how it be achieved? *Forum on Public Policy Online*, No. 2. https://forumonpublicpolicy.com/summer08papers/archivesummer08/bohlin.pdf. Accessed 1 Dec 2023.

Carrington, D. (2024, January 9). 2023 smashes record for world's hottest year by huge margin. *The Guardian.*

Collier, A. (1994). *Critical realism.* Verso.

Conlon, E. (2008). The new engineer: Between employability and social responsibility. *European Journal of Engineering Education, 33*(2), 151–159.

Conlon, E. (2013). Broadening engineering education: Bringing the community in. *Science and Engineering Ethics, 19*(4), 1589–1594.

Conlon, E. (2019). Prisoners of the capitalist machine: Captivity and the corporate engineer. In S. H. Christensen, B. Delahousse, C. Didier, M. Meganck, & M. Murphy (Eds.), *The engineering- business nexus: Symbiosis, tension and co-evolution*. Springer Nature.

Conlon, E. (2022). Engineering ethics, social theory and how we might do better! In S. H. Christensen, A. Buch, E. Conlon, C. Didier, C. Mitcham, & M. Murphy (Eds.), *Engineering, social sciences, and the humanities* (Philosophy of engineering and technology) (Vol. 42). Springer.

DFHERIS. (2023). *Statement of Strategy 2023–25.* Department of Further and Higher Education, Research, Innovation and Science.

Engineers Ireland. (2021). *Accrediation Criteria.* Engineers Ireland.

Fellenz, M. R. (2016). Forming the professional self: *Bildung* and the ontological perspective on professional education and development. *Educational Philosophy and Theory, 48*, 267–283.
Fenner, R. A. (2023). *Reflections on engineering sustainability: Progress, pedagogy, principles and practice*. EESD2023 Keynote. CSU.
Freire, P. (1972). *Pedagogy of the oppressed*. Penguin Books.
Huckle, J. (2017). Powerful geographical knowledge is critical knowledge underpinned by critical realism. *International Research in Gepgraphical and Environmental Education, 28*(1), 70–84.
Huckle, J., & Wals, A. E. J. (2015). The UN decade of education for sustainable development: Business as usual in the end. *Environmental Education Research, 21*(3), 491–505.
IBEC. (2018). *Future ready: Improving graduate employability skills*. IBEC.
Jamison, A. (2013). *The making of green engineers*. Morgan and Claypool.
Jamison, A., Kolmos, A., & Holgaard, J. E. (2014). Hybrid learning: An integrative approach to engineering education. *Journal of Engineering Education, 103*, 253–273.
Johnson, D., & Wetmore, J. M. (2021). *Technology and society*. MIT Press.
Karwat, D. M. A., Eagle, W. E., Wooldridge, M. S., & Princent, T. E. (2015). Activist engineering: Changing engineering practice by deploying practice. *Science and Engineering Ethics, 21*(1), 227–239.
Lindsay, E. D., Hadgraft, R. G., Boyle, F., & Ulseth, R. (2023). Disrupting engineering education. In *International Handbook of Engineering Education Research* (pp. 115–133). Routledge.
Martin, D. (2020). *Towards a sociotechnical reconfiguration of engineering and an education for ethics: A critical realist investigation into the patterns of education and accreditation of ethics in engineering programmes in Ireland*. PhD Thesis, TU Dublin.
Martin, D. A., Conlon, E., & Bowe, B. (2021). A multi-level review of engineering ethics education: Towards a socio-technical orientation of engineering education for ethics. *Science and Engineering Ethics, 27*, 60.
McClaren, P. (2001). Foreward. In P. Allman (Ed.), *Critical education against global capitalism*. Bergin and Garvey.
McGeown, C., & Barry, J. (2023). Agents of (un)sustainability: Democratising universities for the planetary crisis. *Frontiers in Sustainability, 4*, 116662.
Moody, K. (2021). Workers of the world: growth, change, and rebellion. *New Politics, XVIII*(2) https://newpol.org/issue_post/workers-of-the-world-%E2%80%A8growth-change-and-rebellion/. Accessed 20 Feb 2024
Mooney, A. (2023, June 1). Earth has been pushed past its safe limited for humans, scientists say. *Irish Times*.
Moore, R. (2004). *Education and society: Issues and explanations in the sociology of education*. Polity Press.
Moore, R. (2013). Social realism and the problem of knowledge in the sociology of education. *British Journal of Sociology of Education, 34*(3), 333–353.
Mulder, K. F. (2019). Is progress to sustainability committed engineers stalling? *Journal of Modern Education Review, 9*(1), 23–28.
Murphy, M., Christensen, S., & Conlon, E. (2022). Social justice at an Irish practice-based university: In or out of place? In S. H. Christensen, A. Buch, E. Conlon, C. Didier, C. Mitcham, & M. Murphy (Eds.), *Engineering, social sciences, and the humanities: Have their conversations come of age?* (Philosophy of Engineering and Technology Series) (Vol. 42). Springer.
Nicolaou, I., Conlon, E., & Bowe, B. (2018). Into the deep: The role of paradigms in understanding engineering education for sustainable development. *Irish Journal of Social, Economic and Environmental Sustainability., 1*(2), 23–41.
Noble, D. (1977). *America by design*. Oxford University Press.
Oxfam. (2023) *Richest 1% bag nearly twice as much wealth as the rest of the world put together over the past two years*. https://www.oxfam.org/en/press-releases/richest-1-bag-nearly-twice-

much-wealth-rest-world-put-together-over-past-two-years#:~:text=The%20richest%201%20percent%20. Accessed 20 Feb 2024.

Pavlidis, P. (2013). The ideal of education and the emancipation of labour. *Policy Futures in Education, 11*(3), 299–312.

Pearce, F., & Tombs, S. (1998). *Toxic capitalism*. Ashgate.

Priestly, M. (2011). Whatever happened to curriculum theory? *Pedagogy, Culture & Society, 19*(2), 221–237.

Ramnath, A., Bix, A. S., Winberg, C., Pevkur, A., & Conlon, E. (2023). The historically global forming of engineering education. In *International handbook of engineering education research* (pp. 9–29). Routledge.

Restad, F. (2019). Revisioning the Fifth Element. Can critical realism reconcile competence and *Bildung* for a more sustainable twenty-first-century education? *Journal of Critical Realism, 18*(4), 402–419.

Scatamburlo-D'Annibale, V., Brown, B. A., & McLaren, P. (2018). Marx and the philosophy of praxis. In P. Smeyers (Ed.), *International handbook of philosophy of education*. Springer International Handbooks of Education.

Schaull, R. (1972). Forward. In P. Freire (Ed.), *Pedagogy of the oppressed*. Penguin Books.

Seely, B. (2005). Patterns in the history of engineering education reform: A brief essay. In D. C. Washington (Ed.), *Educating the engineer of 2020: Adapting engineering education to the new century* (pp. 114–130). National Academies Press.

Sjöström, J., Frerics, N., Zuin, V. G., & Eilks, I. (2017). Use of the concept of *Bildung* in the international science education literature, its potential, and implications for teaching and learning. *Studies in Science Education, 53*(2), 165–192.

Slaughter, J. (2002). The political economy of curriculum making in American universities. In S. Brint (Ed.), *The future of the city of the intellect* (pp. 260–293). Stanford University Press.

Small, R. (2014). *Karl Marx-The revolutionary as educator*. Springer.

Spring, J. (1975). *A primer of libertarian education*. Black Rose Books.

Sterling, S. (2004). Higher education, sustainability and the role of systematic learning. In P. B. Corcoran & A. E. J. Wals (Eds.), *Higher education and the challenge of sustainability*. Kluwer.

Sterling, S. (2011). Transformative learning and sustainability: Sketching the conceptual ground. *Learning and Teaching in Higher Education, 5*, 17–33.

Torres, C. A., & Yan, L. (2022). Paulo Freire and the state-of-the-art of the international journal of lifelong education. *International Journal of Lifelong Education, 41*(6), 651–665.

Wheelahan, L. (2010). *Why knowledge matter in the curriculum*. Routledge.

Wheelahan, L. (2015). Not just skills: What a focus on knowledge means for vocational education. *Journal of Curriculum Studies, 47*(6), 750–762.

Winner, L. (1990). Engineering ethics and political imagination. In P. Durbin (Ed.), *Broad and narrow interpretations of philosophy of technology: Philosophy and technology* (Vol. 7). Kluwer.

Wintour, P. (2023, September 18). Western leaders defend slow progress to end global inequality as UN summit starts. *The Guardian*.

Xavier, P., Wint, N., & Orbaek White, G. (2022). A snapshot of how 'social' considerations are currently being interpreted and addressed within engineering education and accreditation. In S. H. Christensen, A. Buch, E. Conlon, C. Didier, C. Mitcham, & M. Murphy (Eds.), *Engineering, social sciences, and the humanities* (Philosophy of engineering and technology) (Vol. 42). Springer.

Young, M. (2008). From constructivism to realism in the sociology of the curriculum. *Review of Research in Education, 32*(1), 1–28.

Eddie Conlon is a sociologist and until 2024 worked at the Technological University of Dublin. He worked with engineering students for almost 20 years to help them explore the relationship between technology, engineering, and society. His interests are in the sociology of work but recently has published, and supervised graduate students, in areas such as engineering ethics, the social position of engineers, the integration of sustainability into engineering education and the use of case studies in teaching ethics. He has published in Science and Engineering Ethics and the European Journal of Engineering Education.

Chapter 8
Bildung, and How That Concept Sits with Traditional Notions of Teaching Students "How to Think" Like Engineers

Karl Palmås

Abstract Starting from the proposition that higher education is a matter of "learning how to think", this chapter investigates the meaning of *Bildung* in the context of engineering education. Following recent calls for a re-evaluation of classical *Bildung*—more specifically for a turn to a more politically-oriented critical-reflexive version of *Bildung*—the chapter studies one earlier instance of a re-evaluation of classical *Bildung*. By exploring how Erik Gustaf Geijer and Carl Jonas Love Almqvist challenged dominant conceptions of *Bildung* in Sweden during the 1830s and 1840s, the chapter argues that the classical *Bildung* ideal of forming original human beings who can think for themselves remains a valid objective of higher education. As such, the chapter suggests that recent, more politically-oriented versions of *Bildung* jeopardize the classical *Bildung* ideal of intellectual autonomy—an ideal which holds particular value in engineering education.

Keywords Liberal education · Classical *Bildung* · David Foster Wallace · Erik Gustaf Geijer · Carl Jonas Love Almqvist

8.1 Introduction: Learning "How to Think"

It is sometimes held that higher education is about teaching students "how to think". Indeed, I remember hearing this repeated to me when I was an engineering student, having ploughed through the initial maths-intensive phase of the program. Doing engineering, a teacher told me, is not so much about learning to solve this or that

K. Palmås (✉)
Department of Technology Management and Economics, Chalmers University of Technology, Gothenburg, Sweden
e-mail: karl.palmas@chalmers.se

A. Buch, S. H. Christensen (eds.), *Bildung for Engineering Education and Practice*, Philosophy of Engineering and Technology 49,
https://doi.org/10.1007/978-3-031-86581-7_8

equation. Rather, it is about learning to think in a specific manner. As such, each particular, specific solution of a mathematical problem is a drill exercise that hones one's skills in more general problem-solving. As such, engineering education is really about instilling a habit of mind; about teaching students to engage with the world in a specific problem-solving manner. The same *modus operandum*, over and over: Breaking large and complicated problems into manageable components, solving each lower-level problem one by one, thus deriving a higher-level solution.

This description of engineering education is hardly unique: In his recent monograph about *Bildung* in engineering education, Jacobsson (2019, p. 44) suggests that the problem-solving trope constitutes the most pervasive notion of what it means to think like an engineer. Similarly, when the UK Royal Academy of Engineering describes what it means to "think like an engineer", problem-finding and problem-solving are promoted as key "engineering habits of mind" (Royal Academy of Engineering, 2014). Again, from this perspective, engineering education is a matter of teaching students to think in such problem-solving terms.

However, the proposition that higher education is about teaching students "how to think" is not confined to engineering education. The idea is as prevalent in the Anglo-American tradition of liberal arts education. Due to the fact that they do not—like technical universities—present their *raison d'être* in terms of educating for a specific profession, liberal arts institutions are more strongly wedded to the abstract purpose of "learning how to think" (Harford, 2022; Schwartz, 2015). As such,

> the goal of liberal education cannot simply be the acquisition of content knowledge, but the development of thinking skills, and more specifically the skills of thinking insightfully, critically, and creatively. (Hadzigeorgiou, 2015, p. 263)

American novelist David Foster Wallace comments on this proposition in his famous 2005 commencement speech to the graduating class at the prestigious liberal arts institution Kenyon College. In the meditation subsequently published as *This is Water*, he even suggests that the idea of teaching students how to think is the "single most pervasive cliché" in the liberal arts commencement speech genre. (Foster Wallace, 2009) For him, the problem is not only that it is a hackneyed trope: It is also somewhat insulting to students. The young minds that he was addressing were already capable thinkers before they arrived at the highly selective Ohio-based college.

Nevertheless, Foster Wallace concludes, the cliché of learning "how to think" actually *does* contain an element of truth, but this truth requires some qualification. For one, a liberal arts education does offer the student a distinct set of tools with which to understand the world. This includes what Foster Wallace calls the "standard liberal arts analysis", which prompts the student to accept that "the exact same experience can mean two totally different things to two different people, given those people's two different belief templates and two different ways of constructing meaning from experience". This understanding of the world breeds tolerance, and an appreciation of diversity of belief—two virtues that are key to liberal education. Similarly, Schwartz (2015) demystifies the notion of learning how to think by spelling out a series of intellectual virtues that should be fostered by liberal education. These include honesty about one's failings; fair-mindedness; humility and willingness to seek help; good listening; perspective-taking and empathy. Schwartz are keen to bill them precisely as virtues—a term that implies moral imperatives—which are distinct from cognitive skills such as quantitative abilities, conceptual flexibility, analytical acumen, and expressive clarity. Similarly, when the American Association for the Advancement of Science promotes liberal education in the sciences, it describes it in the following terms:

> Ideally, a liberal education produces persons who are open-minded and free from provincialism, dogma, preconception, and ideology; conscious of their opinions and judgments; reflective of their actions; and aware of their place in the social and natural worlds. (AAAS, 1990: xi)

Still, Foster Wallace refrains from positing that this "standard" liberal education habit of mind is the be-all and end-all of liberal education. The value of such studies cannot be reduced to individuals learning how to think in tolerant and open-minded ways. "Learning how to think", he tells the graduating class, really means "learning how to exercise some control" over what to think about. It is about learning to *choose* what you think about, as you find yourself in the "day-to-day trenches of adult existence".

For him, then, the value of a liberal education is deeply personal, as it may assist in coping with the emptiness of contemporary life. Through the liberal arts, students may escape feelings of boredom and loneliness—major themes of Foster Wallace's novels—but only through exercising control over their thoughts. Actively choosing what to think about in the context of the contemporary world is crucial, especially as the standard "socially conscious" liberal arts criticality may sometimes increase one's sense of emptiness. In short, Foster Wallace seems to suggest that the virtues and habits of thought cultivated by a liberal education are noble pursuits—but ultimately students must be capable of *transcending* "standard" procedures, "standard" habits of thoughts, and become autonomous subjects.

This reading of Foster Wallace's argument—one that emphasizes the cultivating of individuals capable of thinking for themselves—places it squarely within a long-standing discussion on *Bildung*, as conceived in the continental European setting as well as Anglo-American liberal education tradition. This idea will occur throughout this chapter, which will argue that this aspect of classical *Bildung* should remain central in current discussion on how to interpret the notion of *Bildung* in contemporary engineering education.

The argument runs as follows. The next section will review central themes in the literature on *Bildung* in engineering education, focusing on the five different conceptions of *Bildung* outlined by Sjöström et al. (2017), as well as the kind of updated, politically oriented critical-reflexive *Bildung* that they themselves promote. It will also introduce Erik Gustaf Geijer and Carl Jonas Love Almqvist, whose work was central to how the notion of *Bildung* was introduced and re-interpreted in the Swedish educational setting during the nineteenth century. Neither Geijer nor Almqvist were associated with technical universities, but their concerns—what is the place of classical *Bildung* in the context of new social challenges?—overlaps with the issues at stake in this argument.

The subsequent section (8.3) goes into the specifics of Geijer and Almqvist's understanding of the place of classical *Bildung* in a modernizing society, emphasizing their point of not lapsing into static modes of reading that merely mimic the classics. The section then transposes their work into the issue of teaching engineering students: In contrast to the close readings associated with the humanities, the context of the engineering education requires creative readings that make the problems inherited from the classics come alive in the context of present concerns. More generally, Geijer and Almqvist's work underscores the importance of promoting creativity rather than fidelity in engaging with the classics. This imperative is somewhat at odds with recent interpretations of a critical-reflexive *Bildung* that suggest that STEM education should be aligned with social critique and the agendas of societal challenges. Much like liberal education is about more than repeating the "standard liberal arts analysis", *Bildung* in engineering education cannot be reduced to teaching students the "standard" procedures for social critiques of engineering, or standard critiques of technology in general.

The chapter ends with a concluding discussion (8.4) that joins the threads of the argument. Combining the arguments of Foster Wallace, Geijer, Almqvist and others, the argument ends with an attempt to describe and name that "excess" faculty that we seek to promote when fostering *Bildung* among engineering students.

8.2 Shifting Conceptions of *Bildung*

Thus far in this chapter, two versions of *Bildung* have been discussed: The classical *Bildung* tradition of continental Europe, and the liberal education tradition of the Anglosphere. In a useful systematic review of international literature on the concept, Sjöström et al. (2017) list three further versions of *Bildung*: The Scandinavian folk-*Bildung* tradition, the United States-based tradition of democratic education, and the more recent critical-hermeneutic version of *Bildung*. This section will describe these in further detail, and then briefly describe where the work of Geijer and Almqvist fit into this taxonomy.

8.2.1 Five Versions of Bildung

Classical *Bildung* has its roots in continental Europe and tends to be associated with Vilhelm von Humboldt. In this original statement of the term, *Bildung* is 'a process of individualization where humans develop their personality through studies and reflections in a diverse, harmonious, and unique way, and thus become a human original rather than a copy of others' (Burman, 2014, p. 127, cited in Sjöström et al., 2017, p. 169). Thus, the term emphasizes a process of personal development, which serves a purpose that transcends any professional or instrumental ends. This sentiment is inherited from German romanticism. There is value in seeking knowledge for knowledge's own sake because the end of *Bildung* is self-realization. Not only does this end transcend professional imperatives—it also trumps ends like the common good, or the unity of the state. (Beiser, 2003, p. 91) This is why autonomy is crucial: The Humboldtian university should be autonomous—free from outside interference—and the end of the education is the autonomous, fully developed human being.

The tradition of classical *Bildung* also inherits another sentiment from German romanticism—the primacy of the humanities and fine arts. For one, there is Humboldt's own preoccupation with language. For him, *Bildung* is about overcoming a fundamental alienation from the external world (von Humboldt, 2000, p. 59), and thus about establishing a connection between the self and the world. This connection can only ever be structured as language, ranging from everyday speech to the adoption of alternative worldviews through the learning of foreign languages. (von Humboldt, 1963) However, there is also the influence of the German romantics' focus on *Bildung* through art, not least through Friedrich Schiller's aesthetic education. This idea emerged in response to what the romantics saw as a failed enlightenment: The thinkers of the enlightenment had sought to explain (*aufklären*) the world though reason, but by spreading their ready-made concepts to the public they had contributed to revolutionary chaos. Their approach was flawed, Schiller suggested, because it did not encourage individuals to think for themselves, or thinking spontaneously, since the enlightenment philosophers had already done the thinking for them. The public had become passive recipients of knowledge, not autonomous individuals, and could not translate the philosophers' ideals into reasoned action. (Beiser, 2003, p. 94) Schiller thus proposed that aesthetic experience of the beautiful could develop sensibilities that compel the individual to act in reasoned ways. The inspirational power of art would fill the moral void left after enlightenment's demolishing of religious authority. In this way, aesthetic education serves as a higher order mode of learning—rather than building formal knowledge, it shapes the character of autonomous human beings.

As hinted in the previous section, the Anglo-American tradition of liberal education is shares many of the traits of classical *Bildung*. It too emphasizes the formation of individuals who can think for themselves, and the idea of seeking knowledge for its own sake. However, given that its ties are not as strong to the romantic tradition of continental Europe, it has traditionally not placed as strong of a focus on the humanities. Rather, liberal educational programs are generalist in nature, encompassing the sciences as well as the humanities and social sciences. In studying a broad range of fields—mathematics, science, literature, fine arts, history, and philosophy—liberal arts students are attuned to different "forms of knowledge". (Hirst, 1972) As such, the liberal education "trains the seeing of connections" (Kallenberg, 2015, p. 133) across knowledge domains.

Aside from the above, there are two versions of *Bildung* that are more directly oriented towards the education of all citizens. In Scandinavia, there is the folk-*Bildung* tradition, which emerges in the late nineteenth century. As a part of the social-democrat ambition to build an inclusive society, it is an endeavor to provide education for the whole people, outside of the confines of academia. In the United States, the democratic education—which is associated with pragmatist philosopher John Dewey—represents a similar ambition.

Finally, Sjöström et al. (2017) traces a critical-hermeneutic *Bildung* tradition, that starts with Hans-Georg Gadamer (1900–2002) and Paul Ricœur (1913–2005), runs via Wolfgang Klafki (1927–2016), and on to present-day thinkers. As a continuation of this tradition, they propose a critical-reflexive *Bildung*, updated for the present time. This involves a stronger socio-political orientation, enabling it to "function as a bridge between traditional liberal education and activism-oriented education", promoting "constructive participation in society, and solidarity towards persons limited in the competence of self-determination and participation". (Sjöström et al., 2017, p. 180).

This more politicized version of *Bildung* is deemed useful in today's "late modern risk society with all its socio-scientific and technological challenges". The present time is one that places demands on educational institutions, and one may wonder how the ideals of classical *Bildung* fare in the context of such challenges. Let us now turn to an earlier time in history—to the 1830s and 1840s—when similar anxieties about social challenges prompted similar re-evaluations of classical *Bildung*.

8.2.2 Geijer and Almqvist's Challenge to Classical **Bildung**

By European standards, Sweden was a late bloomer in terms of democratization and modernization. Major advances in political participation and socio-economic welfare set in a few decades into the twentieth century. However, in the early to mid-nineteenth century, Swedish intellectuals were closely monitoring the onset of modernity that was raging on the European continent.

Erik Gustaf Geijer—historian, poet, statesman and Rector at Uppsala University—was one of those who saw the modern groundswell coming and adjusted his political orientation accordingly. He had spent the early 1800s as a conservative intellectual, championing the proposition that only nobility and clerics should learn how to think. Here, he was heavily influenced by the German romantic idea of *Bildung*, and its concomitant focus on the humanities. (Geijer, 1928a [1810]; Landquist, 1924) In the late 1830s, he would then reject this position, announcing his defection from conservatism to liberalism, and spearheading an educational reform that saw the state assuming responsibility for offering *Bildung* to the broader citizenry of the emerging middle class. (Geijer, 1928b [1838]) This would later pave the way for the development of the Swedish folk-*Bildung* tradition. Crucially, his departure from his previous views also represented a departure from the idea that humanities should hold a privileged position when cultivating *Bildung*.

Geijer's experience is emblematic of a specific period in history, in which a conservative, feudal state of affairs was crushed by a modernity that—paraphrasing Marshall Berman—not only brought "adventure, power, joy, growth, transformation of ourselves and the world" but also threatened "to destroy everything we have, everything we know, everything we are". (Berman, 1982) Geijer personified the shift from conservatism to liberalism, but also an abandonment of a specific reading of *Bildung*, and a shift towards a new one.

The liberal reform of the educational system that Geijer promoted was also supported by the novelist and educationalist Carl Jonas Love Almqvist. Their joint engagement on the issue constituted a challenge to what classical *Bildung* had become in Swedish public life. In the next section, this chapter will explore the specifics of this challenge to classical *Bildung*. At this point in the argument, however, it is worth noting two things. First: Our moment in history—the "late modern risk society with all its socio-scientific and technological challenges" (Sjöström et al., 2017, p. 180)—is not unique in generating calls for a revision of what *Bildung* entails. Second: Then, as now, classical *Bildung* was challenged by progressives who suggested that new times call for new interpretations of the term. However, as we shall see in the next section, their proposed revision of the term differed significantly from today's calls for a more activist- and societal participation-oriented interpretation of *Bildung*.

8.3 Learning from Geijer and Almqvist

This section will first explore the specifics of Geijer and Almqvist's critique of what classical *Bildung* had become in the 1830s. It will then transpose some of their points to current concerns with how to cultivate *Bildung* in contemporary engineering education, focusing specifically on how to read classics, and how to inherit the legacy of classical *Bildung* today.

8.3.1 From Romanticism to Realism

As hinted above, Geijer defected from conservatism to liberalism, and this shift also amounted to a new interpretation of *Bildung*. This transformation can be described as a shift from romanticism to realism, and this shift affected not only Geijer. In the intellectual circles in Stockholm and Uppsala of the 1830s and 1840s, romanticism was in decline, and there was a "realist urge to get away from empty abstractions". (Norberg, 1944, p. 67) Realism is, however, a particularly apt description of Geijer's position. His turn to liberalism can be understood as emerging from an experience of being a conservative "mugged by reality".[1] In light of the new social, economic, and technical realities he observed in Europe, he realized that the feudal structures

[1] This is a paraphrase—a near-reversal of Irving Kristol's famous saying that a neoconservative is a liberal mugged by reality.

(and the conservative views that defended them) were about to crumble. This caused him some concern—he feared that modernity may spell doom—but he nevertheless felt compelled to face what he saw as new facts about the world. His defection from conservatism to liberalism caused an outcry, and there is no coincidence that it was announced in the context of a debate on education.

Both Geijer and Almqvist concluded that they were living in a time when the affirmation of the individual was inevitable. This understanding was, in part, a result of a reading of the social circumstances, but it was equally a reflection of their inspiration from classical *Bildung*; the celebration of self-actualization. They both referred to the "personality principle"—a phrase coined by Geijer that implied that we become fully formed individuals in the meeting with others. Such meetings must, however, be meetings among equals. Thus, Geijer—like Tocqueville, his major source of inspiration—held that the gravitation towards equality was both desirable and inevitable. The personality principle contained both political and pedagogical dimensions. First, the primacy of the individual and the inevitability of equality implied that the remnants of the feudal system of government had to be abolished. However, the personality principle connected the question of rights—*one person, one vote*—with the question of *Bildung*—*education for all*. (Ullman, 2014, p. 46).

This is where the critique of classical *Bildung* surfaced. On one level, this critique was directed at the problem of what constitutes proper education for political representatives. On this matter, Almqvist provided a scathing critique, stating that some of the elements of classical *Bildung*—such as knowledge in Greek or Latin—was as useless as knowledge in heraldry. (Almqvist, 1995 [1844–1845], p. 393) Instead, the new times required knowledge in governance, relating to military matters, as well as fundamentals of finance and economics. (p. 394) Almqvist railed against how the nobility and clergy referred to classical *Bildung* when legitimating their power—especially as their *Bildung* constituted a shallow one. The only thing that distinguished the rulers from the ruled was the fact that the rulers could use French, German, English or Latin to say what the ruled could say in Swedish. The nobility and clergy made use of hackneyed sayings that "sound more than they contain" but held little relevance for the political matters at hand. Their modes of thought were convoluted, and they lacked the capacity to continue a train of thought to their logical end. (p. 395).

However, Almqvist and Geijer's intervention also had effects beyond the issue of how political representatives were to be educated. Their efforts contributed to the establishment of a public school system for all, aimed at the education of citizens into self-government and self-actualization. This development also implied a toppling of the hegemonic status of Latin course of study (consisting of classical languages) as the education for the ruling classes. Latin was thus complemented by a new course of study—fittingly titled Real—which focused on mathematics and natural sciences.

For the purposes of this chapter, it is worth examining the details of Almqvist's account of how the Latin course of study had failed the original ideals of classical *Bildung*. In his critique of the supposedly "learned" ruling classes, one can trace his animosity towards how the study of classical languages had been reduced to an empty mimicking, devoid of creativity. This ran counter to the imperatives of forming autonomous human beings who could think for themselves. According to the personality principle, the point of education must always be for individuals to seek their own originality. When engaging with the classics, he suggested, the challenge is to avoid shallow repetition of dead sayings, and instead probe the deeper meanings that such classics contain. Indeed, as Ullman (2014, pp. 53–54) suggests, his position is in alignment with the more suggestion by Donald Broady (1984)—that classical *Bildung* cannot be inherited through mere "aping". The only way to inherit it is through constructive engagement with the experiences and ideas of the classics. The chapter will now move on to examine what import these historical developments may have on contemporary engineering education.

8.3.2 Inheriting the Ideals of Classical Bildung *Today*

As suggested above, the debate on classical *Bildung* in early- to mid-nineteenth century Sweden is illuminating when placed alongside contemporary discussion on how to inherit the ideals of such *Bildung* in engineering education. Then, like now, a sense of societal dislocation provides the impetus for a re-evaluation of what *Bildung* really means. Beyond these similarities, there are however further things to take away from Geijer and Almqvist. One concerns the specific question of how to read the classics, another the broader question of how to foster constructive participation for engineering students in society.

In the North American context, there are recent attempts to formulate what a "liberal studies in engineering" program would look like. (Bucciarelli & Drew, 2015) For instance, this could imply adding a broader understanding of the sociopolitical context of engineering, through "schooling in certain 'fundamentals' […] in political philosophy and social theory, for example, Plato, Aristotle, Machiavelli, Hobbes, Locke, Rousseau, Marx, and Tocqueville". (Bucciarelli & Drew, 2015, p. 117) In response to this proposal, Donna Riley (2015) warns against shallow appropriations of the liberal arts approach, where it becomes reduced to "a haphazard smattering of superficial exposures". She describes such cringe-worthy practices in the following manner:

> Add an element of history here, a writing assignment there, an ethics case study every blue moon and presto! Liberal education. The most dangerous aspect is that engineering faculty dilettantes can mischaracterize or reduce entire genres of knowledge, damn the consequences. (Riley, 2015, p. 139)

This is, of course, a valid concern. It is wise for faculty of technical universities not to overstep the mark and mislabel their approaches. Established course titles like "Technology and Society" (van der Vleuten et al., 2017, p. 157) are perhaps more useful for describing the teaching activities that seek to instill *Bildung* in engineering curricula—which, after all, are more instrumentalist in nature than either liberal education or the humanities. Still, it is worth remembering that such activities at technical universities are not *meant* to act as torchbearers of the liberal education or classical *Bildung* traditions.

Furthermore, as suggested by Almqvist, one must resist the tendency for classical *Bildung* to gravitate towards non-creative repetition, towards the shallow mimicking of dead sentences. This constitutes a betrayal of the ideals of classical *Bildung*. Again, the challenge when reading classic texts—be it Plato, Machiavelli, or Hobbes—is to move from shallow repetition to the deeper meanings. In the end, the timeless character of such texts—their capacity to speak to people across ages and contexts—is precisely what makes them classics. This implies that they can indeed speak to today's engineering students, provided their reading of them is a creative one. Such readings require that the teacher manages to convey some deeper meaning of the text that remains valid in the context of the students' future life. The means by which this meaning is acquired are, nevertheless, of a lesser importance: As compared to the humanities, there is less need for back-to-back readings of complete works, or for readings in original language.

In other words, the creative readings that may be productive in the context of engineering education must not be confused with the *close* readings that are the mainstay of the humanities. Such close readings remain crucial for safeguarding our collective ability to inherit from classic texts. Riley (2015) is correct in arguing that engineering faculty members may be well served with a bit of humility towards the rich tradition that they are poaching from. Nevertheless, that should not stop them from persisting in their attempts to assign classic texts to their students. After all, in doing so, they breathe new life into such texts.

The question of how to inherit the ideals of classical *Bildung* is also at stake in the recent proposition by Sjöström et al. (2017) that *Bildung* today should prompt students to engage in "constructive participation in society". Again, one may compare this re-evaluation of the meaning of *Bildung* to that of Geijer and Almqvist. While both proposals for re-evaluation point to an urgency created by social circumstances, there are clear differences between the two.

Sjöström et al. (2017)—along with other scholars—point to the need for a critical-reflexive form of *Bildung* that forms "a bridge between traditional liberal education and activism-oriented education", tying the *Bildung* agenda to specific political challenges, notably social and ecological justice. (Sjöström et al., 2017, p. 182) This position also includes the fostering of critical literacy, in which reflection breeds action. In this view, critical *consciousness* is invariably tied to critical *agency* (Ebenezer, 2013): Critical literacy implies a disruption of the commonplace, an interrogation of multiple viewpoints, a focus on socio-political issues, tied to political action and promoting social justice. (Lewinson et al., 2002) The kind of citizenship that is at stake is a critical one, and this includes the fostering of political emotions. For Ruitenberg (2009, p. 277, cited in Sjöström et.al., 2017, p. 180) this may—in accordance with the political theory of Chantal Mouffe—imply "the ability to feel anger on behalf of injustices committed against those in less powerful social positions".

In contrast, Geijer and Almqvist's re-evaluation of classical *Bildung* concerned the domains of knowledge that are to be privileged: Away from a sole focus on classical languages, to more focus on social sciences and natural sciences. It also concerned the methods of engaging with the classics, with a focus on deep understanding rather than shallow regurgitation. The objective of these revisions was, however, to stay true to the original ideals of classical *Bildung*, inherited from the romantics—the cultivation of unique human beings, which possess the capacity to think for themselves. This was, in turn, into their view of citizenship: The very point of democracy is to release the originality of each singular individual.

Here, the position of Geijer and Almqvist is at odds with the more recent formulation of a critical-reflexive *Bildung*. In short, the highly prescriptive orientation of the latter approach leaves little room for autonomous individuals who are to think for themselves. There is a risk that critical-reflexive *Bildung*—at least when operationalized into everyday teaching—moves beyond the idea of teaching students *how* to think, instead prescribing *what* to think. There is also a risk that students call teachers out to prescribe what to think *about*, as well as how to *act* and *feel*. What to think about? Social and ecological justice. How to act? In accordance with such justice movements. How to feel? Anger.

Thus, much of Almqvist's critique of the shallow Latin-based *Bildung* of his time could be directed toward such prescriptive tendencies of contemporary critical-reflexive *Bildung*. True—for Geijer and Almqvist, the issue of cultivating responsible citizens was at stake, but this process could not be traded off against the process of forging individuals' ability to think for themselves. After all, shallow regurgitation was precisely what Almqvist objected to.

These propositions may be transposed onto *Bildung*-oriented teaching in contemporary engineering education contexts. When studying classics like Plato, Machiavelli, or Hobbes, there is ample space *not* to prescribe what students must take-away from the class. The teacher may encourage a particular way of understanding the text—for instance, that one of the concerns of Plato is the place of expertise in decision-making. However, the question of what to think about Plato or his critics, what to feel about such views, or whether the discussions in class should prompt political action is best left to the students themselves. Yet, as stated before, classics are classics by virtue of their ability to invite new thoughts and feelings—to suggest new potential waypoints for student's continued travels. In the end, it is up to them which waypoints to follow as they develop their own original character. In the end, an all too prescriptive approach places teachers in a position perilously close to the false impression that *Bildung* is a readymade bundle of ideas, rather than a process of personal formation.

As such, classical *Bildung* implies that education is to produce something more—or perhaps something else—than whatever their teachers convey to students. This problem will be pursued in the concluding discussion.

8.4 Concluding Discussion: The Excess of Education

This chapter has argued that the current discussion on a re-evaluation of classical *Bildung* may be well served by reviewing one such earlier re-evaluation—the challenge posed by Geijer and Almqvist in the 1830s and 1840s. For one, the work of Almqvist is useful for thinking about how to engage with classic texts in engineering education. For engineering students, *creative* readings are more productive than the *close* readings conducted within the humanities. However, Almqvist is also interesting considering recent proposals for a more politicized form of critical-reflexive *Bildung*. It is worth noting that his sentiments are not grounds for dismissing critical-reflexive *Bildung* because it challenges some ideal of an "objective" technical education: The political ramifications of engineering practice should indeed be conveyed to students. The point is, rather, that students must use their education to figure out their own political or ethical positions.

This reading of Almqvist partly dovetails with Hadzigeorgiou's (2015) concerns that 'science education as socio-political action' (SEASPA) approaches downplay "the importance of knowledge for its own sake". (Hadzigeorgiou, 2015, p. 259) However, while Hadzigeorgiou's disquiet is primarily directed towards a neglect of the "personal/aesthetic dimension of science"—encompassing a sense of awe, mystery, and wonder—the argument above is not solely relating to personal matters. Given the long history of engineers running errands for the powers that be (Berner, 1981; Riley, 2008), a focus on teaching engineering students to think for themselves can be understood as a safeguard against future abuses of power. Intellectual autonomy not only serves the private human being—it may also serve the wider community. If personal originality of mind is the final objective of education, then the *Bildung* efforts of engineering schools should steer clear of becoming too prescriptive on matters socio-political. If not, education fosters subservience rather than autonomy.

At this point, we may return to Foster Wallace's meditation on learning how to think—or rather, on how to learn to *choose* what to think about. His message can be read as one that celebrates the very same autonomy described above. Yes, the "standard liberal arts analysis" taught during the education may be useful, but in the end, the true gift is the ability to choose to *transcend* that formal "take-away" from the education. This begs the question of what an education should offer, if its value really resides in that which is not formally taught. This paradox goes all the way back to the German romantics. Thus,

> there was nothing more important to the romantics than *Bildung*, the education of humanity. This was the central theme and goal of their ethics, aesthetics, and politics. But, from a more practical perspective, there seems to be *nothing less important* to the romantics than education. When it comes to concrete suggestions about how to educate humanity—about what specific institutional arrangements are to be made—the romantics fell silent. (Beiser, 2003, p. 105, italics added)

In his own way, Foster Wallace seems to be making a similar point. So is Swedish educationalist Ellen Kay, who famously argued that *Bildung* is that which remains once you have forgotten what you have learnt. The classical *Bildung* ideal of forging autonomous thought presents something like a reversed version of Bataille's (1991) "accursed share": It is not the excess of an education is accursed; it is *only* the excess that is *not* subject to disposal. Formal knowledge must be squandered for the "formed" human being to come into view.

Another way to approach this paradox is to ask what critical thinking really implies. As Bruno Latour (2004) has suggested, it all too often implies reducing something more to something less, when in fact criticality should be generative. Here, he engages in Alan Turing's discussion on intelligence among humans and computers. Turing suggests that most of the minds that one encounters among humans are "sub-critical", which means that an "idea presented to such a mind will on average give rise to less than one idea in reply". However, a "smallish proportion" of human minds are "supercritical":

> An idea presented to such a mind may give rise to a whole "theory" consisting of secondary, tertiary and more remote ideas. (Turing, 1950, p. 454)

Turing suggests that animal minds are always sub-critical, but the question is whether "a machine be made to be super-critical?". Turing's analogy is, of course, sourced from the idea of critical mass in fission: A sub-critical mass is not sufficient for setting off a chain reaction—nothing "more" can emerge from the sub-critical. For Latour, on his part, is more interested in human minds, and in what the notion of "super-criticality" means for our understanding of the term "critique". What would it mean if by "critical" we always mean a habit of mind that yields something more, not less? A habit of mind that is generative, not reductive? A fostering of that mode of critical thinking among engineering students may be the proper way to inherit the tradition of classical *Bildung*.

In summary, there are several overlapping figures of thought that may capture the meaning of classical *Bildung*. They all point towards the idea of *Bildung* as the excess of education: For Almqvist and Geijer, *Bildung* must transcend the empty mimicking of classical languages, and instead contribute to the formation of unique, individual "personalities". For Foster Wallace, a liberal arts education should provoke something more than the "standard liberal arts analysis" and enable students to choose for themselves what to think about. The German romantics were passionate about *Bildung* as the general objective of education, but agnostic about the practical means of formal schooling. Similarly, Kay suggested that while you may forget the formal knowledge acquired during an education, *Bildung* is that which remains. In this way, *Bildung* can be understood as an inverted version of Bataille's "accursed share". Finally, one may understand *Bildung* in relation to Latour's and Turing's simile about "super-critical" intelligence: *Bildung* should forge minds that go beyond simply reducing the world to the pre-existing knowledge imparted through education.

While all these figures of thought are evocative, the question still remains: How should engineering education be organized in order to promote such *Bildung*? Here, it is important to not retreat to the above-mentioned agnosticism about educational practices displayed by the German romantics. However, one must also recognize that there are no educational silver bullets or quick fixes on this matter. The notion of classical *Bildung*, understood as the excess of education, *does* rest on a fundamental paradox. Any educational effort to promote *Bildung* must start from recognizing this vexing fact. Indeed, the problem of how to teach something as elusive as *Bildung* looms large in the above-mentioned discussions on the place of the humanities and liberal arts in engineering education.

Aside from such scholarly discussions, however, it is also productive to invite engineering students to consider this paradox. Simply by discussing the wider objectives of higher education—the fact that an engineering education involves *both* instrumentalist knowledge *and* an excess of *Bildung*—students may become more cognizant of the *Bildung* aspect of their education. Thus, they may form their own ambitions for the years they spend at university. They may recognize that while acquiring formal knowledge, they are also in a process of forming their personality, becoming unique human beings. Placing their attention on this fact is important: A process of becoming easily goes unnoticed by the person being subject to it. Further, in becoming attuned to Ellen Kay's insight into the ephemeral nature of formal knowledge and the lasting impact of *Bildung*, students may consider the longer-term benefits of acquiring an education, beyond landing their first job.

There is one further merit in presenting engineering education as an inheritor of the ideals of classical *Bildung*. In so doing, it is possible to clarify the rules of engagement between teachers and students. Again, the point of the education is for students to transcend whatever the teacher tells them. As proposed by Hannah Arendt, the role of the teacher is to merely present "what the world is like", but "not instruct in the art of living". (Arendt, 2006, p. 195) The teacher can only represent the present and the past world; the future belongs to the students.

> Our hope always hangs on the new which every generation brings; but precisely because we can base our hope only on this, we destroy everything if we so try to control the new that we, the old, can dictate how it will look. (p. 192)

This is one way of understanding the stakes of classical *Bildung* in the context of contemporary engineering education. We teachers are educating the caretakers of a future world we know very little about. All we know is that the formal technical knowledge of today will become obsolete. The best we can do is to assist in the formation of unique, whole human beings to whom we can safely pass the torch.

References

AAAS. (1990). *The liberal art of science: Agenda for action*. American Association for the Advancement of Science.

Almqvist, C.J.L. (1995 [1844–1845]). *Monografi*. Svenska Vitterhetssamfundet.

Arendt, H. (2006). *Between past and future*. Penguin Books.

Bataille, G. (1991). *The accursed share: An essay on general economy* (Consumption) (Vol. 1). Zone Books.

Beiser, F. C. (2003). *The romantic imperative: The concept of early German Romanticism*. Harvard University Press.

Berman, M. (1982). *All that is solid melts into air: The experience of modernity*. Simon and Schuster.
Berner, B. (1981). *Teknikens värld: Teknisk förändring och ingenjörsarbete i svensk industri*. Arkiv förlag.
Broady, D. (1984). Om bildning och konsten att ärva. *Kritisk utbildningstidskrift, VIII*, 35–36.
Bucciarelli, L., & Drew, D. (2015). Liberal studies in engineering – A design plan. *Engineering Studies, 7*(2–3), 103–122.
Ebenezer, J. (2013). Social justice pedagogy for all science learners. *Studies in Science Education, 49*, 252–264.
Foster Wallace, D. (2009). *This is water: Some thoughts, delivered on a significant occasion, about living a compassionate life*. Little Brown.
Geijer, E. G. (1928a [1810]). Om inbillningsgåvan. In *Samlade Skrifter* (Vol. 1). Norstedt och Söner.
Geijer, E. G. (1928b [1838]). Tal vid invigning av nya katedralskolehuset i Uppsala d. 21 okt. 1837. In *Samlade Skrifter* (Vol. 8). Norstedt och Söner.
Hadzigeorgiou, Y. (2015). A critique of science education as sociopolitical action from the perspective of liberal education. *Science & Education, 24: April*, 259–280.
Harford, T. (2022, June 10). Learning how to think well involves hearts as well as minds. *Financial Times*.
Hirst, P. (1972). Liberal education and the nature of knowledge. In R. Dearden, P. Hirst, & R. Peters (Eds.), *Education and the development of reason* (pp. 391–414). Routledge & Kegan Paul.
Jacobsson, P. (2019). *Den bildade ingenjören: Bildning, utbildning och teknik*. Studentlitteratur.
Kallenberg, B. (2015). Liberal arts is more than 'perspective'. *Engineering Studies, 7*(2–3), 132–134.
Landquist, J. (1924). *Erik Gustav Geijer: Hans levnad och verk*. Norstedt och Söner.
Latour, B. (2004). Why has critique run out of steam? From matters of fact to matters of concern. *Critical Inquiry, 30*(2), 225–248.
Lewinson, M., Flint, A. S., & Van Sluys, K. (2002). Taking on critical literacy: The journey of newcomers and novices. *Language Arts, 79*, 382–392.
Norberg, E. (1944). *Geijers väg från romantik till realism*. Svenska kyrkans diakonistyrelses bokförlag.
Riley, D. (2008). *Engineering and social justice*. Morgan and Claypool.
Riley, D. (2015). Facepalms and cringes: Liberal education misapprehended. *Engineering Studies, 7*(2–3), 138–141.
Royal Academy of Engineering. (2014, May). *Thinking like an engineer: Implications for the education system*. Report for the Royal Academy of Engineering Standing Committee for Education and Training.
Ruitenberg, C. W. (2009). Educating political adversaries: Chantal Mouffe and radical democratic citizenship education. *Studies in Philosophy and Education, 28*, 269–281.
Schwartz, B. (2015, June 18). What 'learning how to think' really means. *The Chronicle of Higher Education*.
Sjöström, J., Frerichs, N., Zuin, V. G., & Eilks, I. (2017). Use of the concept of *Bildung* in the international science education literature, its potential, and implications for teaching and learning. *Studies in Science Education, 53*(2), 165–192.
Turing, A. M. (1950). Computing machinery and intelligence. *Mind LIX, 236*, 433–460.
Ullman, A. (2014). Rektor C.J.L. Almqvist och personlighetsprincipen. *Nordic Journal of Educational History, 1*, 2.
van der Vleuten, E., Oldenziel, R., & Davids, M. (2017). *Engineering the future, understanding the past: A social history of technology*. Amsterdam University Press.
von Humboldt, W. (1963). Schriften zur Sprachphilosophie. In A. Flitner & K. Giel (Eds.), *Werke IV*. WBG.
von Humboldt, W. (2000). Theory of *Bildung*. In I. Westbury, S. Hopmann, & K. Riquarts (Eds.), *Teaching as a reflective practice: The German didaktik tradition* (pp. 57–61). Erlbaum.

Karl Palmås Associate Professor at the Division of Science, Technology and Society, Chalmers University of Technology in Gothenburg, Sweden. Originally trained as an engineer, he holds a PhD in Sociology from the London School of Economics and Political Science. His teaching involves training the professional judgment of future engineers by learning from classics in social thought. His research entails sociological studies of design activities—be it in architecture, automotive industry, nanotechnology, or spatial planning—and his most recent English-language book is *The Corruption of Co-Design: Political and Social Conflicts in Participatory Design Thinking* (Routledge, 2023), co-authored with Otto von Busch.

Chapter 9
Appropriateness and Value: Cultivating *Bildung* Mindsets Amongst Undergraduate Engineers in a Third-Year Design Course

Sarah Appelhans and Alan Cheville

Abstract *Bildung* as a flexible concept offers multiple perspectives to integrate aspects of becoming into education. This chapter describes experiments in implementing ideas of *Bildung* into an engineering classroom. At a liberal arts university in the United States, an Electrical & Computer Engineering department has undertaken the task of teaching undergraduate engineers how to evaluate the social impacts of technology. Contemporary North American liberal arts education is broadly associated with the concept of *Bildung* in its emphasis on holistic development of individuals who are capable of independent thought, which is often achieved through a firm grounding in the humanities and social sciences. Although this connects with an expansive notion of *Bildung* as an active relationship between oneself and the world, in this work, we also emphasize the emancipatory spirit of critical-reflexive forms of *Bildung* through our engagement with integration of socio-technical knowledge in engineering design.

Consistent with the recommended uses of *Bildung* in STEM, the degree program is shifting from an outcomes-based assessment model to a holistic curriculum that incorporates socio-political knowledge and supports students' individual development through the use of reflexive activities. One of the crucial interventions in this curriculum is during the third-year design course. In this class, technical problems take a backseat to socio-political ones. Students select a complex societal problem that they will intervene in with a technological solution. Throughout multiple design iterations, they are encouraged to continuously evaluate the "appropriateness" and "value" of the product for the given context. Students integrate design decisions with socio-political contexts through the creation of graphical representations, a method embedded within engineering epistemologies. The course concludes with an ePortfolio assignment, in which students reflect upon their personal journey in

S. Appelhans (✉) · A. Cheville
Lafayette College, Easton, PA, USA

Department of Electrical and Computer Engineering, Bucknell University, Lewisburg, PA, USA
e-mail: appelhas@lafayette.edu; alan.cheville@bucknell.edu

A. Buch, S. H. Christensen (eds.), *Bildung for Engineering Education and Practice*, Philosophy of Engineering and Technology 49,
https://doi.org/10.1007/978-3-031-86581-7_9

the course, the insights they have learned, and the importance it may have going forward in their careers.

The chapter authors are a senior electrical engineering faculty member and an anthropologist who was embedded in the course for observation. Participant observation methods were employed in the design classes over two semesters. Qualitative codes were generated to characterize faculty's desired learning goals and field notes were analyzed using this codebook. A set of vignettes have been selected to demonstrate the successes and challenges of using *Bildung* approaches in engineering contexts. Early successes include students' acquired understanding of the importance of social context, iterative design, and critical feedback. Challenges include students' resistance to take on "just a social science project" and failures of imagination regarding social context outside of their immediate experiences, ethics, and what might go wrong in the use of their designs. The chapter concludes with a set of practice-derived recommendations on how to support the development of *Bildung* capacities in engineering courses, and the ways these may support students as they continue to develop over their lifetimes and careers in engineering.

Keywords *Bildung* · Capabilities approach · Engineering

9.1 Introduction

This chapter analyzes a single engineering design class at a United States (US) liberal arts university through the lens of *Bildung*. The Central/Northern European *Bildung* philosophy envisions education as an iterative and reflective process of self-development that contributes to societal improvement. Professors serve as metaphorical 'guide stars', pointing in a desired direction of travel while leaving the details of the journey flexible to enable student agency and growth. *Bildung* perspectives contradict normative engineering education programs that prepare students for jobs by focusing narrowly on education as a vehicle for the development of specific skills and knowledge. Such transactional views are increasingly enmeshed in US discourses about the purposes of higher education due to the precipitous rise of tuition costs in the last decades (College Board, 2015). *Bildung* also contests dominant epistemologies that frame engineering disciplines as a form of applied science which is abstractable from, and thus unconcerned with, social impacts. The emphasis on engineering science, particularly in the middle two years of undergraduate programs (Kotys-Schwartz et al., 2010), diverts students' interest away from the larger effects engineering and technology have on society (Cech, 2014).

Despite the fact we teach at a liberal arts institution that ostensibly espouses *Bildung* principles (Sjostrom & Eilks, 2018), by their junior year our engineering students have become enculturated to expect that courses will teach linear approaches to technical knowledge abstracted from its messy societal impacts. This chapter discusses a curricular intervention designed to begin to shift student mindsets from these siloed, vocational approaches to engineering education and more towards a

Bildung approach, which centers individual agency, moral development, and responsibility to society. Here we seek to address three questions relevant to the intersectional space of engineering, *Bildung*, and the liberal arts:

1. What might a shift from siloed technical content towards a *Bildung* approach look like at the undergraduate level where students are still in a developmental stage?
2. How can such a shift be practically implemented in existing engineering programs given practical and epistemological constraints?
3. What are the challenges students face in adjusting to a course organized around *Bildung* principles and how do they respond?

9.2 *Bildung*, US Liberal Arts Education, and Engineering

Bildung, in a broad sense, refers to the values and processes of an individual growing in knowledge and skills in a cultural milieu they both draw from and contribute to. Sjostrom & Eilks (2018) identified five threads of *Bildung*: (1) Classical *Bildung*—the ability of individuals to pursue knowledge free from state or market influence; (2) Liberal *Bildung*—the concept of education as cultivating independent thought and life-long learning; (3) Scandinavian-folk *Bildung*—the promotion of individual action and responsibility for the good of society; (4) Democratic *Bildung*—the preparation of individuals to exercise their participation in democracy; and (5) Critical-Hermeneutical *Bildung*—an emphasis on moral development of individuals through critical-reflection, as opposed to education for the development of skills. These threads have emerged over time as different regions responded to different societal and cultural norms. Although *Bildung* is not a part of the cultural milieu in the authors' US liberal arts institutions, *Bildung* shares many values with the liberal arts. The mission and values of intellectual exploration and independent thought espoused by American liberal arts institutions resonate in particular with the Classical and Liberal branches of *Bildung*.

It is the Scandinavian and Critical-Hermeneutical branches of *Bildung*, however, which provide opportunities to move beyond dominant epistemologies of engineering as an "applied science" and towards a model which acknowledges engineers' moral responsibilities and societal impacts. Goldman (1991) has argued that engineering is not, in fact, an "applied science" since it adopts a very different philosophical stance than science. In brief, while science tends to align itself with necessity, i.e. things that must be true, engineering encompasses both necessity and contingency, things that may or may not be true depending on context. Put simplistically, science seeks universal answers while engineering concerns itself with local truth in particular contexts. Another way to describe engineering is that its practitioners must move in both directions on the 'vector of abstraction' that points from particular examples to universal truths (Krupczak & Bassett, 2012). While science abstracts from example and evidence to theory, engineering also goes the other way in utilizing theories as guides for specific, contingent actions in particular situations. Mitcham (2014) highlights that the ability of technology to affect society through engineering gives engineers an often unrecognized role as change agents.

Despite the impact of engineering on society, its characterization as an applied science has been argued by Goldman (1991) to contribute to the 'captivity of engineering' since it constitutes a refusal to recognize engineering as a separate discipline with its own intellectually distinct tradition. Additionally, engineering is captive to social forces that seek to use technology for their own ends. As Goldman somewhat provocatively states, "*These are not two separate captivities. They are complementary expressions of a fundamental usurpation of the intellectual and social dimensions of engineering as an autonomous discipline*." Engineers themselves adopt narratives that frame engineering as applied science (Pawley, 2009) with a focus on solving problems and making things.

Our interest in the intersection of engineering and *Bildung* stems from *Bildung*'s elements of independence or autonomy, culture, and some notion of moral good encompassing the individual's capacity *and* responsibility for action. To hold the values embedded in *Bildung* is to have the capability to act rightly (Varela, 1999) whether or not action occurs. The action-oriented nature of engineering work puts engineering—the discipline that reshapes society through technology–at the 'sharp, pointy end' of *Bildung*. While this is not a typical characterization of the *Bildung* concept, Taylor (2017) points out that *Bildung* has always been contingent and adapts to changing circumstances and times. The tensions and questions of *Bildung* we seek to address in this chapter lie at the intersection of engineers' epistemological emphasis on action with the enacted awareness of acting morally and reflectively for societal benefit.

In this chapter we explore the concept of *Bildung* as a framework that supports both individual agency and moral and social responsibility as core components of engineering education. A significant body of work in engineering education has similarly sought to mitigate engineering's captivity by making it more connected to society and seeking to adopt a more critical praxis. Cech (2014) showed that engineering programs not only neglect to teach the larger contexts of engineering, but discourage students' interest in big picture questions over time. Work to connect engineers has repositioned engineering in social justice (Baillie et al., 2012) and liberal studies (Bucciarelli & Drew, 2015) frameworks, embedded social justice issues into technical engineering content (Riley, 2012), provided more holistic frameworks for learning engineering (Vanasupa et al., 2009; Vanasupa & Barabino, 2022), and examined engineering ethics (Vesilind, 2010) and design (Kallenberg, 2013) through societal lenses among many others. Previous books in this series have highlighted the need for, and value of, conversations between engineering, social sciences, and humanities (Christensen, 2022). Framed in an engineering context, each of these efforts develops a 'toolset' of approaches designed to shift the center of gravity of engineering education (Froyd et al., 2012) towards more inclusion of societal issues inherent to *Bildung*.

The next section shifts to a curricular intervention designed to adapt two core tenets of *Bildung*, individual agency and moral responsibility, into an engineering course. We seek to explore the messy and often imperfect realities of classroom and curricular interventions and understand how such changes impact students.

9.3 Methods

To understand the generative tension of liberal arts and engineering as it intersects with *Bildung*'s key tenets, an ethnographer was embedded in the Junior Design class, described in the next section, for two consecutive semesters. Ethnographic methods enable close observation of students grappling with social realities and their own developmental journey at the everyday classroom level.

9.3.1 Participants and Data Collection

Data was collected over two semesters of a third year Design course in the 2021-22 academic year. Permission to observe was granted by the professor teaching the course, but all students were given an opportunity to opt out if they wished. Pseudonyms were assigned to all students at the beginning of the semester and all research materials were anonymized. All class sessions were attended by an ethnographer who took detailed field notes throughout both semesters. Participant observation is a valuable method for capturing everyday, ordinary experiences. Ethnography is particularly useful for examining the values and meanings within communities that are unspoken and often subconscious (Emerson et al., 2011; Geertz, 1973; Van Maanen, 1988) and is an important tool for documenting processes of change. In the larger study of which this is a part, we are documenting change at multiple scales: (1) the evolution of a department as it experiments with reframing its values, (2) the growth of professors as they expand their domains of expertise, and (3) the growth of students who are learning and integrating new knowledge, skills, and mindsets.

The ethnographer was present for all class sessions, often sitting with students as they worked on in-class assignments. She kept a record of collaborations between students on their teams, student uptake of (and occasional resistance to) the professor's core themes, moments of difficulty with particular concepts, and important breakthroughs in understanding. She also regularly checked in with the professor to capture his perspective on the class's progress. According to best practices, a concerted effort was made to maintain clear distinctions between observed phenomenon and the ethnographer's own opinions and interpretations in recorded field notes (Bernard, 2011; Emerson et al., 2011). The ethnographer had access to all student assignments, including reading reflections, design representations, peer reviews, and design reports. Of particular importance to this study were the ePortfolio assignments, which took the place of a final exam. ePortfolios revealed students' insights into what they felt was most important in the class and their perceptions of their own progress. These personal reflections were triangulated with field notes from the observer's perspective for holistic insights into students' initial exposure to *Bildung* principles.

9.3.2 Data Analysis

Field notes were coded using a combination of open and structured coding techniques (Elo & Kyngäs, 2008; Hsieh & Shannon, 2005). First pass Descriptive Coding (Saldana, 2013) was used to sort data into core themes. These themes were structured according to an earlier survey of faculty, who had identified several learning goals they were trying to achieve. However, if data did not easily fit into one of these themes, a new category was created. For example, "Autonomy" was not an outcome that faculty had expressly identified, but observation data revealed that this was something professors consistently emphasized in class. A set of memos was generated following the completion of coding for each theme to synthesize what faculty claim to value, what they actually teach in class, and what students appear to learn (from the ethnographers' point of view). These memos were circulated to the faculty for feedback.

Second-pass data analysis involved triangulating field notes, professor's learning goals, and student reflections on their own learning. The ePortfolio assignments were coded using open coding to allow themes to emerge from the students' perspective. A return to observation data revealed important moments in students' learning experiences, the meanings of which were explored in reflective memos (Emerson et al., 2011). A final comparison with the professor's intended goals for the course informs the degree to which students acquired the intended skills and mindsets. Table 9.1 summarizes our findings in relation to *Bildung* principles of individual agency, moral development, and responsibility to society.

Table 9.1 Summary of the effectiveness of *Bildung* concepts in junior design

Bildung concept	Communicated through	Effectiveness
Individual Agency: "The Hero's Journey"		
Autonomy	The Gesture	Positive impact, work in progress
Narrative Contextualization	ePortfolios	Positive impact, work in progress
Moral Development & Societal Responsibility: "Appropriateness and Value"		
Nonlinear, Iterative Design	Iterative Class Structure	Positive impact, goal achieved
Problem Framing	NABC Chart, System Diagram	Mixed results, work in progress
Social Context & Critical Thinking	System Diagram, Ethics Prompts	Mixed results, more work needed

9.3.3 Author Positionality

The authors of this paper are the ethnographer (first author) and the professor teaching the Junior Design course (second author). The first author ("Prof. A"), at the time of observation, was a postdoctoral scholar who has degrees in Cultural Anthropology and Mechanical Engineering. While students often treated her with the respect generally afforded to professors, she got to know them at a more personal level and maintained some degree of separation between herself and the professors (i.e. she did not grade their assignments). The second author ("Prof. C") is a Professor of Electrical and Computer Engineering who has previously served as department chair in the ECE department, and as such, has considerable *gravitas* amongst the students. They are two members of a six-person team on an NSF Revolutionizing Engineering and Computer Science Department's *Revolutionizing Engineering Departments* grant.

9.4 Classroom Context

On the first day of the third-year design class, 'Junior Design', the students surmised almost immediately that this class would be … different. "While your previous design courses have taught you about fabrication and testing," Prof. C said, "in this class you will decide what should be designed in the first place." A tall man with a philosophical bent, the professor was a formidable figure at the front of the classroom, drawing upon his authority to impress upon his audience the importance of ethics in design. A quote appeared on the monitors at each group of tables around the room:

> On the floor of Jove's palace there stand two urns, the one filled with evil gifts, and the other with good ones … he to whom Jove sends none but evil gifts will be pointed at by the finger of scorn, the hand of famine will pursue him to the ends of the world, and he will go up and down the face of the earth, respected neither by gods nor men.—Homer, The Illiad

"Who is Jove?" Prof. C asked. A long silence drew out as students avoided eye contact, unused to being asked to interpret classical literature. "We are all Jove," he told them. "Engineers create the world and as such, we have a certain responsibility. If you create only bad things, you will continue to create evil in the world." The students stared back silently, seriously. Although at this point they had taken several design classes, in this one, the training wheels would come off and they would be plunged into the messiness of compromise and uncertainty that characterizes engineering design.

9.4.1 Appropriateness and Value

Junior Design, the fourth of six design courses, is taught in an Electrical and Computer Engineering (ECE) department at a private liberal arts college in the Mid-Atlantic region of the United States. Earlier courses introduced students to elements of design. In ECE the design process is defined by eight different principles (Cheville et al., 2021): help people and the planet, embrace the context, choose useful functions, design transparently, build responsibly, improve performance, act professionally, and create value. By the time students reach Junior Design their projects have focused on function, design, building, and performance. Junior Design emphasizes the lenses of value, professionalism, people & planet, and context – the "social" aspects of design. The final two courses are the capstone sequence taken in the students' final year, which integrates all eight design lenses.

Junior Design—taught in a project-based, flipped classroom format—is organized around three iterations of the design process. The first third of the semester is devoted entirely to problem-framing (Bucciarelli, 1996; Cross, 2006; Dym et al., 2005) with small teams of students selecting a UN Sustainable Development Goal (UNSDG) (United Nations, 2021) and identifying some aspect amenable to a technological intervention. The middle third of the course asks students to develop a minimum viable product (MVP) (Ries, 2011). In the last third, students iterate their design to create a contextually relevant device or service.

Junior Design is the first design course in which students are able to negotiate and define their own project topic. It is the class which most embodies *Bildung* ideals through the development of students' intellectual autonomy and attention to social contexts in design, which are communicated to students through core concepts of "appropriateness" and "value". "Appropriateness" calls attention to the importance of tailoring designs to particular contexts while "value" emphasizes that engineering designs must be useful. In contrast to a narrowly economic sense of value, Prof. C. takes a broad sense of the term, emphasizing the value of the design in improving the lives of users and communities. Throughout the course, when students drifted into technical tunnel-vision, the language of appropriateness and value refocused students' attention on the social aspects of design.

9.4.2 Representations

To apply appropriateness and value, students created visual representations as an aid to the design process. Representations are common in engineering design (Barner et al., 2022; Bucciarelli, 1996; Kallenberg, 2013) and used to improve learning in STEM disciplines (Moore et al., 2013). In this class, they helped students explore the full scope of their problems, synthesize complex ideas, and represent them in a simple and understandable way. Although the social sciences often use writing and reflection to help students work through the complexities of social phenomena, representations capture these complexities in a concise, visual format that is more appealing to engineers.

Representations included technical schematics, like block diagrams, but also problem-framing diagrams. The Needs-Approach-Benefits-Competition (NABC) Chart (Carlson & Wilmot, 2006) helped students articulate a clear problem and value statement, and a System Map (Stroh, 2015) captured the context and impacts of technical interventions. Although students tended to treat representations like homework assignments that were completed and forgotten, the professor maintained that their representations *are* their designs; they should be revisited and updated with each iteration.

9.4.3 The Hero's Journey

The culminating assignment was a reflective digital ePortfolio based on Joseph Campbell's (1973) "the hero's journey". Students narrated their personal experience in the course, including personal challenges and victories along the way by curating content generated in the class such as representations and photographs of their final product. ePortfolios and other reflective activities help students synthesize the lessons they have learned and negotiate their identities in relation to the engineering profession (Chen, 2009; Moon, 1999; Yancey, 2016; Yancey & Rhodes, 2019). These portfolios are particularly useful for cultivating *Bildung* mindsets; they encourage students to see their education as a journey within which they have some degree of agency and cultivate a sense of curiosity about the world and their place in it.

9.5 *Bildung* Themes Emerging from Ethnographic Data

Intended and actual course learning outcomes often differ. Ethnographic observations revealed that there were some areas where embedding *Bildung* principles were successful and others where more attention was needed. This section is organized around the two core *Bildung* principles we have previously identified: (1) individual agency and (2) moral development and responsibility to society. Each of the subsections below explores emergent themes from classroom observations (see Table 9.1). Within each theme, we offer three perspectives: (a) Prof. C's communication and intentions, (b) Prof. A's observation of classroom activities, and (c) the students' ePortfolio reflections.

9.5.1 Individual Agency

Junior Design was the first course in which students are entirely on their own in choosing and defining a problem. Students were encouraged to explore areas related to their own interests and then reflect on their progress in their culminating activity, the ePortfolio. The ePortfolio is what Prof. C. referred to in class as "narrative contextualization", which enables students to recognize their struggles, their achievements, and their values.

9.5.1.1 Autonomy

Developing autonomy was an important learning outcome in Junior Design. Skimming through the UN Sustainability Goals (UNSDG) and the NAE Grand Challenges, students identified a problem they had some intrinsic interest in addressing and conducted research on potential solutions. Prof. C. offered guidance regarding techniques and problem scoping during the design reviews, but to support autonomy, final decisions were always left to the students.

There is a gesture Prof. C makes when he feels the students are relying upon him too much for direction. Holding his hands together, as if holding a bowl of soup, he looks down at the problem the student has brought to him. Having looked at the problem, he wordlessly hands the problem back to the student. Although Prof. C is sympathetic to the students' struggle, he believes that an appropriate amount of struggle is critical to their learning and development of agency.

The initial reaction from students was often confusion and frustration. For example, a team that chose to investigate Virtual Reality-based education struggled to learn the VR design software. Sam[1] wrote in his ePortfolio: "Since we went with Virtual Reality, it was a topic that … well, our professor did not have a Ph.D. in virtual reality. I think other groups could really lean on the professor when they ran into trouble, but after a certain point, we were on our own."

However, over time students learned to be resourceful in finding their own solutions. Many were grateful for the experience and recognized their own capacity for resilience. Daniel reflected, "I think that one of the most important things I learned that will apply later in life, is learning how to learn difficult things in an unstructured environment. The need to continue learning will be paramount in anyone's career." This is exactly the lesson Prof. C hoped students would take from this experience.

[1] All student names are pseudonyms.

9.5.1.2 Narrative Contextualization

Similarly, narrative contextualization was a concept that students initially struggled with and later learned to appreciate. Students who took the course in the fall semester did not grasp the narrative structure, treating the portfolio as a technical report despite Prof. C's warning, "If it looks like a data sheet, it's a failure." In the spring semester, the professor adjusted the template for the portfolio to explicitly follow the stages of the Hero's Journey (Campbell, 1973), which generated results that were more self-reflective.

One student, Luke, expressed strong resistance toward narrative itself. In one assignment, Prof. C. asked students to retell their problem statements as if they were stories about real people. Luke was visibly uncomfortable during the activity. "I don't like narrative," he said flatly in his report to the class.

Undeterred by Luke's defiance, Prof. C asked the rest of the class to describe the value of a narrative account. One student said he thought the narrative style was more emotional, and as a result, more convincing. Another thought the narrative was easier to remember. Frustrated, Luke interjected, "Narratives make me feel manipulated." Patiently, the professor agreed. "Narratives are powerful. Especially when you are asking for money or persuading people to do what you want, it is important to be truthful."

Although students were unpracticed and resistant to narratives in the beginning, most made an honest attempt. Of the twenty-five students in both classes, only four students refused to engage in the assignment entirely. Even Luke, whose portfolio explored his frustration with the class, concluded:

> I think that's the issue with me, I look at things too broadly, I see the entire problem that is being faced with its complexities and all that, and decide at the start that it's not even worth it to try. But that's not a reasonable way to look at the world, and to a certain extent this class has helped lessen that worldview, taught me to look at things in smaller portions, in sizes I can tackle.

In this sense, the goal of the ePortfolio has been accomplished. As Luke reflected on his negative experience, he acquired self-knowledge and reframed the experience to reflect his positive growth.

9.5.2 Moral Development and Responsibility to Society

Prior to this course, most engineering students had not yet engaged deeply with moral and ethical questions. To situate these concepts within the culture of engineering they had encountered thus far, such issues were framed through concepts of "appropriateness" and "value". The pedagogical tools that Prof. C. used to facilitate these new concepts included: (1) multiple iterations with design review feedback, (2) the NABC Chart, (3) the System Diagram, and (4) ethics prompt cards. These tools were intended to help students create designs that were useful to real people.

9.5.2.1 Non-Linear, Iterative Design

Perhaps the most successful lesson from the Junior Design class was the framing of design as an iterative process. On the first day of class, the professor outlined the contrasts between "engineering science" and "engineering design". Engineering design problems cannot be solved linearly, he explained. In design, "appropriateness" and "value" will vary based on the context. Lacking clear "right" or "wrong" solutions, it was better to make a first attempt, get feedback, and then improve the design over numerous iterations.

After each design review, students were encouraged to "pivot", or make substantial changes in direction, in response to feedback from the professors. At one design review, Prof. C. pushed a team to clearly state the *value* of their product, a sensor to monitor pollutants in rivers and oceans. "The existing water stations are expensive," Antonio explained. "It's mostly scientific researchers who do water testing. We want it to be accessible to the general public."

Affirming the value of raising public awareness about pollution, Prof. C. mused, "I wonder if this would make a good public science project." The team enthusiastically pivoted to an educational context where K-12 students would track a "Polluto" sensor journeying through U.S. waterways. To produce a design *appropriate* for education, they sacrificed precision for cost-effectiveness, and provided additional instructions for students to build their own "Polluto" device.

By the end of the class, the majority of students had internalized the importance of iteration. In their final ePortfolios, nearly all reflected on the importance of being able to "pivot". Trevor reflected, "While initial ideas may seem promising, there are moments of uncertainty, drastic changes, abandoned ideas, but all of them are necessary in getting closer to that final design with major flaws worked out." Notably, this passage reflects a positive orientation toward failure, which was echoed by many students in the class.

9.5.2.2 Problem Framing

Students floundered in the problem-definition phase. In the first few weeks of class, Prof. C advised students several times to "keep the problem space open for as long as possible." However, there were three common blunders: (1) becoming overwhelmed during initial exploration, (2) jumping straight to the solution before understanding the problem, and (3) struggling to translate problems into representations.

On the day Prof. C introduced System Diagrams, he gave clear instructions: map the problem first, then show how the technology changes the cycle. However, a team designing a smart parking system began with their technology as the centerpiece. Prof. C soon corrected them: start with the problem. Befuddled, the students abandoned the System Diagram and complained amongst themselves about student parking. Felix grumbled that overcrowding prevented freshmen from having cars on campus. Bill bemoaned how crowded the parking lots were on the North side, which hosted several residence halls.

Using their complaints to inform the System Diagram, Prof. A sketched some new loops, identifying multiple users who had different interactions with the system: freshmen, students on the North Side of campus, etc. The students nodded politely. However, a week later, their completed system diagram remained a rudimentary sketch. As users of the parking lots on campus, students were experts in explaining the overcrowding problem. However, they did not know how to translate their complaints to their representation.

Junior Design students never fully recognized the value of the problem-framing representations. Prof. C had intended for these to be living documents, revised upon each iteration. He hoped that students would use them to drive their decision-making. Unfortunately, students treated them like homework assignments, dutifully dropping them into their design reports, but seldom using them as tools to think with. Although both professors were disappointed that the representations were not as successful as anticipated, they retained hope that this first exposure to problem-framing would prepare them for a second attempt the following year in Senior Design.

9.5.2.3 Social Context and Critical Thinking

The most challenging aspect of this class, for professors and students alike, was connecting meaningfully with societal context and asking students to think critically about their designs. Prof. C again used the language of appropriateness and value to stress the importance of societal context. Quoting Russell Ackoff's (1998) distinction between efficiency and effectiveness, he would warn students, "If you do things efficiently without effectiveness, you only get more wrong."

Rather than using their representations as decision-making tools, as anticipated, students made design decisions in response to ethical prompts that came up in their design reviews. Prof. C. employed a set of ethical prompt cards (Artefact Group, n.d.) to raise questions like, "How could bad actors use this technology for negative purposes?" and "What if the environment was your client?" In reviews, students drew a card at random and attempted to answer.

Students were mostly unable to answer prompts on their own. The Polluto water sensor team, for example, drew a card asking, "How can this be used to help an underserved population?" There are a variety of possible answers to this question, such as, improving data to communities surrounding polluted waterways. The team was flummoxed. Simon ventured, "Well, we are going to focus on kids now, I guess they are underserved." Other teams fell into the lazy solution, "the worst possible scenario is that it doesn't work". Prof. C and Prof. A offered suggestions after students' first attempts faltered, essentially doing the critical thinking on their behalf.

While students struggled with the prompts, they were very influenced by Prof. C's suggestions. Most of the pivots in the class were related to questions Prof. C asked during the design review, such as Polluto's pivot to education.

On a more positive note, in their ePortfolios, over half of the students were pleased to have worked on a project with societal impacts. Chad mirrored Prof C.'s language on efficiency, saying, "If you are negligent and do not put in the effort into the whole design process you will waste your time or even make the problem worse." Several also reported changed understandings of what it means to be an engineer. Theo wrote, "This design class has helped reshape my understanding of what goes into engineering on the less-technical side of the design process. Engineers aim to design a humanitarian solution to a real-world issue affecting the lives of many."

9.6 Discussion

The introduction of *Bildung* principles in Junior Design presents a challenge to dominant engineering epistemologies. The epistemological boundaries become visible in our students' struggle to adapt to this new way of doing engineering. Engineers are accustomed to thinking of themselves as members of a technical, rational domain (Cech, 2013; Faulkner, 2009a, b) often in contrast to disciplines they demarcate as social, emotional, and irrational. Riley (2017) argues that engineers' attachment to mathematical and technical "rigor" in education occludes their exploration of other kinds of rigorous inquiry: complex decision-making, which is needed for ethics; analysis of power, which is needed to understand social change; and clarity of expression, which is needed for communication. Engineering "cultures of disengagement" dissuade engineers from engaging in political conversations that determine the uses of technology (Cech, 2014). Rather than defining their own problems, engineers tend to see themselves as a discipline that solves problems pre-defined by society, industry, or the military (Pawley, 2009) We have argued, with Goldman (Goldman, 1991, 2004), that this is a misunderstanding of engineering epistemology and an abdication of engineers' agency as creators of the built world. *Bildung* presents opportunities to introduce students to new "habits of mind" that encompass complexity, problem-framing, and adaptation (Lucas & Hanson, 2016).

In this class, our students are grappling with their own emerging identities as engineers. On the one hand, there is considerable enthusiasm amongst this generation about solving societal problems. On the other, they are attempting to fit themselves into pre-defined notions of who engineers are and what their responsibilities should be that are formed in prior engineering science courses. Junior Design is a radical departure from the dominant model of engineering they have been enculturated into by their Junior year. The idea that the "social" is indeed part of their purview is a paradigm shift for many.

While *Bildung* concepts guided the course structure and daily interactions in Junior Design, outcomes are always individual and uncertain; there are no 'silver bullets' in student development. Here we step back and evaluate the impact of our practices on the core aspects of *Bildung* from both the ethnographer's and professor's perspectives.

9.6.1 Autonomy

Prof C: Autonomy is not something that is taught but rather it is an offering that has to be designed into a course. Offering students autonomy can be daunting as an instructor since you lose quite a bit of control. Your trust in the fact that students want to learn can't waver or you slip back into 'teaching' mode.

Prof. A: Students initially resisted Prof. C's refusal to usurp their agency. After a lifetime of lecture-based classes, students have become accustomed to being told what to do and parroting back "correct" answers. Indeed, students confessed to Prof. A that lecture classes were a "relief" in comparison to the flipped-classroom model, which demands their active participation. This tendency mirrors engineers' perception that their job is simply to solve problems defined by others. Although students find autonomy frustrating at first, they come to see the value after the struggle comes to a close.

9.6.2 Narrative Contextualization

Prof. C: I've observed many faculty trying to write their students' stories rather than let them emerge, and I wanted to design this course to give students experiences that could be woven into their own stories. You can *learn* engineering through dedication and hard work, but to *become* an engineer you have to write your own story.

Prof. A: The focus on student narrative met with some resistance, which echoes engineers' epistemological commitments to emotional detachment. Although Luke was the most extreme case, most students were inexperienced with this form of writing. Accustomed to presenting their work in a depersonalized way, many students misunderstood the assignment and wrote a report-like portfolio instead. Luke himself was uncomfortable with the ways in which narrative intentionally manipulates emotions. The binary thinking that characterizes engineering science does not adequately equip students to think critically about the story that is being told. Uncomfortable with uncertainty and nuance, Luke was unable to evaluate the truthfulness of narrative, leaving him understandably upset. Fortunately, despite these challenges, students were able to adapt to the narrative style over time.

9.6.3 Moral Development and Social Responsibility

Prof. C: The concept of 'moral injury' emerged from the military to describe the mental damage that occurs when a person observes or participates in events that go against their core values (Shay, 2014). As we grow in understanding of the detrimental effects technology can have on the environment, other species, and social equity I think we risk moral injury to our students if we can't find a way to address the larger issues *Bildung* points us towards. The challenge is how to introduce broader moral considerations to students who are enculturated to engineering ways of knowing.

Prof. A: The technical/social dualism encumbered students' attempts to grapple with ethical and social impacts of design. I suspect that part of the reason the iterative process was so popular amongst the students was that it lowered the pressure for them to get it "right" on the first try. Given their inexperience with the other tools offered to interpret complexity—the System Diagram, the NABC Chart, the ethical prompt cards—iteration is a habit of mind that offers a gentler orientation toward failure and a roadmap to future success.

Prof. C: To me engineering design is liberating since it enables you to make mistakes, to be wrong, and to not feel ashamed in doing so. That freedom comes from iteration, the chance to learn from mistakes and try again.

Prof. A: At the same time, students' lack of exposure to social problems revealed significant weaknesses in their ability to create solutions that will impact society. They struggled to transform complaints into problems that could be solved and they lacked sufficient knowledge, most likely found in the social sciences and humanities, to understand how technologies might be misused. Critical thinking should be a requirement for the engineers who are responsible for building the technologies we all use. However, this was our students' first exposure to applying technology to social problems and it is likely that with sustained emphasis, they may improve over time. Professors should look to the humanities and social sciences for research on teaching critical thinking as this is a routine part of liberal arts education.

Prof C: Giving students a choice of which goals to address early in the course enables them to act in a way that aligns with their own values and asks them to think about what those values are. Also creating representations, which are common in engineering, that serve as lenses on the larger social context helps reaffirm that this is part of the work of engineers. Valuing such work and reaffirming its worth helps integrate it into students' identities.

Prof. A: Prof. C's language of "appropriateness" and "value" also appeared to be a convincing framing for social issues. Students found these concepts more palatable than social justice rhetoric, which many engineers find off-putting due to its political nature (Cech, 2013). Appropriateness and value do not hold the same politically loaded connotations, which helped introduce students to the importance of societal impacts.

Although results were mixed, this experiment with *Bildung* affords opportunities to think about engineering pedagogy in new ways. Lecture-based engineering courses are often conceptualized as vehicles for content-delivery. Courses are deemed "successful" when students have demonstrated mastery over this knowledge. However, the *Bildung* model encourages us to see courses as opportunities for individual transformation. These models are more akin to "communities of practice" (Gee, 2004; Lave & Wenger, 1991) wherein mastery is obtained gradually over time. Success is not necessarily instantaneous, but akin to planting of seeds that will germinate in the future.

9.6.4 Recommendations

Based on the observations shared above several recommendations emerge for aligning engineering courses with *Bildung*. First, courses do not emerge fully formed like Minerva from Jove's head, rather they must evolve over time. We have found that relationships and a guiding philosophy are key to evolution. The plans for Junior Design took place over several years through conversations with colleagues who all contributed ideas, building a sense of shared ownership. *Bildung* served as a philosophical beacon, expressing a set of values in a form flexible enough to be adapted to local needs.

Second, we placed emphasis on ideas and perspectives rather than specific content so course topics can evolve based on student input. The fluid environment this creates is not comfortable for faculty or students, but *Bildung*'s concepts of agency, moral development, and responsibility require personal engagement at levels not typically found in engineering courses.

Third, students need sufficient time to live with any discomfort, so allowing cognitive dissonance to linger and letting learning evolve over weeks or months through iteration is better than seeking linear progress.

Fourth, polysemous words like 'value' and 'appropriateness' can connect students' linear conceptions of engineering to both *Bildung* concepts and ideas from the social sciences. Allowing students to self-negotiate the meaning of such terms aids the process of connecting their work to larger, societal goals.

Finally, it is important to recognize that undergraduate students are actively defining and negotiating an engineering identity (Gee, 2000; Tonso, 2014); for many this identity is still tenuous and fluid. Engineering identities are strongly influenced by early engineering science courses and asking students to reformulate their engineering identity can be stressful. Since identities need both internal and external recognition, it is important for the instructor to validate any shifts of identity towards *Bildung* as accomplishments leading to a full engineering identity.

9.7 Conclusion

This chapter was informed by three questions that informed our exploration of introducing *Bildung* concepts to an engineering design course: (1) How do we shift from teaching siloed, technical content toward a *Bildung* approach?, (2) How might we implement *Bildung* in existing engineering programs, despite practical and epistemological constraints? And (3) How do students respond to this change and what challenges do they face?

On the first point, recognizing that *Bildung* is not a content area but a set of values that define ways to be in the world informed our work, we emphasized creating a less structured learning environment that reflects and supports these values. Shifting from specific content to open-ended goals met with some student resistance. Although they often underperformed on important lessons, over time their outlook on the world became more aligned with the ideas underlying *Bildung*.

Regarding the second question, our experiences in Junior Design show that meeting students where they are and reframing concepts from philosophy and the social sciences to align with the engineering epistemologies–e.g. through diagrammatic representations–helped make *Bildung* concepts relevant to students. Similarly, translating ideas from the social sciences to 'engineering dialect' and framing the concepts in terms of "appropriateness and value" helped students understand that societal impacts are central to engineering work.

The structural arrangement of the six-course 'design thread' also contributed to the success of the Junior Design class. This thread was established intentionally as part of a larger curriculum reform effort (Cheville et al., 2021) to provide students more autonomy in their education. Working through definitions of design with colleagues, developing a common set of eight design principles, and developing the courses was a process of years, but one that allowed *Bildung* concepts to be introduced before the culminating capstone experience in students' last year.

The third question was understood by embedding an ethnographer for the first two iterations of Junior Design as it transitioned to integrating *Bildung* concepts. Students evidenced the linear forms of thinking promoted by engineering science course structures that had well-defined, short-term deliverables; many struggled with the less structured environment of Junior Design. Assignments designed to inform long-term decision making were a challenge to students used to weekly homework. Although some students initially resisted the invitation to engage with social, emotional, and personal orientations toward engineering work, we see evidence of growth in their willingness to accept that these elements are central to determining "appropriateness and value" in design. We expect that students will continue to develop these skills as they continue to the final course in the design thread.

The shift to a *Bildung* philosophy in engineering design requires a reorientation toward engineering pedagogy. Rather than seeing our roles as deliverers of technical content, and measuring 'success' as the retention of concepts, *Bildung* asks us to see learning as a work-in-progress, a slow process of student growth over time. As professors, we are the 'guide stars', pointing the way but otherwise allowing students the freedom some element of choice in direction. By encouraging students to reflect upon their own values and think seriously about their impacts on society, *Bildung* contains the potential to contest the cultures of technical myopia in engineering, and become a more socially-conscious, morally responsible discipline.

This work was funded by the National Science Foundation under EEC-2022271. Any opinions, findings, and conclusions or recommendations expressed in this material are those of the author(s) and do not necessarily reflect the views of the National Science Foundation.

References

Ackoff, R. L. (1998). A systemic view of transformational leadership. *Systemic Practice and Action Research, 11*(1), 23.

Artefact Group. (n.d.). *The tarot cards of tech*. Retrieved June 2, 2023, from https://tarotcardsoftech.artefactgroup.com/

Baillie, C., Pawley, A. L., & Riley, D. (Eds.). (2012). *Engineering and social justice: In the university and beyond*. Purdue University Press. https://doi.org/10.2307/j.ctt6wq5pf

Barner, M. S., Adam Brown, S., Bornasal, F., & Linton, D. (2022). Tangibility of representations in engineering courses and the workplace. *Journal of Engineering Education, 111*(1), 162–184. https://doi.org/10.1002/jee.20439

Bernard, H. R. (2011). *Research methods in anthropology: Qualitative and quantitative methods* (5th ed.). Rowman & Littlefield Publishers.

Bucciarelli, L. L. (1996). *Designing engineers*. The MIT Press.

Bucciarelli, L. L., & Drew, D. E. (2015). Liberal studies in engineering – A design plan. *Engineering Studies, 7*(2–3), 103–122.

Campbell, J. (1973). *Myths to live by*. Penguin Compass.

Carlson, C. R., & Wilmot, W. W. (2006). *Innovation: The five disciplines for creating what customers want*. Crown Business.

Cech, E. A. (2013). The (Mis)Framing of social justice: why ideologies of depoliticization and meritocracy hinder engineers' ability to think about social injustices. *Engineering Education for Social Justice, 67*.

Cech, E. A. (2014). Culture of disengagement in engineering education? *Science Technology Human Values, 39*(1), 42–72. https://doi.org/10.1177/0162243913504305

Chen, H. L. (2009). Using E-portfolios to support lifelong and lifewide learning. In *Electronic portfolios 2.0: Emergent research on implementation and impact* (pp. 29–39). Stylus Publishing.

Cheville, A., Thompson, M. S., & Thomas, S. (2021, July 26). *Adding a "design thread" to electrical and computer engineering degree programs: Motivation, implementation, and evaluation*. 2021 ASEE virtual annual conference content access. https://peer.asee.org/adding-a-design-thread-to-electrical-and-computer-engineering-degree-programs-motivation-implementation-and-evaluation

Christensen, S. H. (Ed.). (2022). *Engineering, social sciences, and the humanities: Have their conversations come of age?* Springer.

College Board. (2015). *Trends in college pricing* (p. 44). Education.

Cross, N. (2006). *Designerly ways of knowing*. Birkhauser.

Dym, C. L., Agogino, A. M., Eris, O., Frey, D. D., & Leifer, L. J. (2005). Engineering design thinking, teaching, and learning. *Journal of Engineering Education, 94*, 103.

Elo, S., & Kyngäs, H. (2008). The qualitative content analysis process. *Journal of Advanced Nursing, 62*(1), 107–115. https://doi.org/10.1111/j.1365-2648.2007.04569.x

Emerson, R. M., Fretz, R. I., & Shaw, L. L. (2011). *Writing Ethnographic Fieldnotes* (2nd ed.). University of Chicago Press.

Faulkner, W. (2009a). Doing gender in engineering workplace cultures. I. Observations from the field. *Engineering Studies, 1*(1), 3–18. https://doi.org/10.1080/19378620902721322

Faulkner, W. (2009b). Doing gender in engineering workplace cultures. II. Gender in/authenticity and the in/visibility paradox. *Engineering Studies, 1*(3), 169–189. https://doi.org/10.1080/19378620903225059

Froyd, J. E., Wankat, P. C., & Smith, K. A. (2012). Five major shifts in 100 years of engineering education. *Proceedings of the IEEE, 100*, 1344–1360.

Gee, J. P. (2000). Identity as an analytic lens for research in education. *Review of Research in Education, 25*, 99–125.

Gee, J. P. (2004). *Situated language and learning*. Routledge.

Geertz, C. (1973). *The interpretation of cultures*. Basic Books.

Goldman, S. L. (1991). The social captivity of engineering. In P. Durbin (Ed.), *Critical perspectives on nonacademic science and engineering*. Lehigh University Press.

Goldman, S. L. (2004). Why we need a philosophy of engineering: A work in progress. *Interdisciplinary Science Reviews, 29*, 163–176.

Hsieh, H.-F., & Shannon, S. E. (2005). Three approaches to qualitative content analysis. *Qualitative Health Research, 15*(9), 1277–1288.

Kallenberg, B. J. (2013). *By design: Ethics, theology, and the practice of engineering*. Cascade Books.

Kotys-Schwartz, D., Knight, D., & Pawlas, G. (2010). First-year and capstone design projects: Is the bookend curriculum approach effective for skill gain. *American society for engineering education annual conference and exposition*.

Krupczak, J., & Bassett, G. (2012). Work in progress: Abstraction as a vector: Distinguishing engineering and science. *Proceedings - Frontiers in Education Conference, FIE*. doi:https://doi.org/10.1109/FIE.2012.6462373.

Lave, J., & Wenger, E. (1991). *Situated learning: Legitimate peripheral participation*. Cambridge University Press.

Lucas, B., & Hanson, J. (2016). Thinking like an engineer: Using engineering habits of mind and signature pedagogies to redesign engineering education. *International Journal of Engineering Pedagogy, 6*(2), 4–13. https://doi.org/10.3991/ijep.v6i2.5366

Mitcham, C. (2014). The true grand challenge for engineering: Self-knowledge. *Issues in Science and Technology, 31*(1).

Moon, J. A. (1999). *Reflection in learning and professional development: Theory and practice*. Kogan Page. http://www.dawsonera.com/depp/reader/protected/external/AbstractView/S9780203822296

Moore, T. J., Miller, R. L., Lesh, R. A., Stohlmann, M. S., & Kim, Y. R. (2013). Modeling in engineering: The role of represntational fluency in students' conceptual understanding. *Journal of Engineering Education, 102*(1), 141–178.

Pawley, A. L. (2009). Universalized narratives: Patterns in how faculty members define "engineering". *Journal of Engineering Education, 98*, 309–319.

Ries, E. (2011). *The lean startup: How today's entrepreneurs use continuous innovation to create radically successful businesses* (1st ed.). Crown Business.

Riley, D. (2012). *Engineering thermodynamics and 21st century energy problems: A textbook companion for student engagement*. Springer International Publishing. https://doi.org/10.1007/978-3-031-79342-4

Riley, D. (2017). Rigor/us: Building boundaries and disciplining diversity with standards of merit. *Engineering Studies, 9*(3), 249–265. https://doi.org/10.1080/19378629.2017.1408631

Saldana, J. (2013). *The coding manual for qualitative researchers*. Sage. http://stevescollection.weebly.com/uploads/1/3/8/6/13866629/saldana_2009_the-coding-manual-for-qualitative-researchers.pdf

Shay, J. (2014). Moral injury. *Psychoanalytic Psychology, 31*(2), 182–191. https://doi.org/10.1037/a0036090

Sjostrom, J., & Eilks, I. (2018). Reconsidering different visions of scientific literacy and science education based on the concept of *Bildung*. In Y. J. Dori, Z. R. Mevarech, & D. R. Baker (Eds.),

Cognition, metacognition, and culture in STEM education: Learning, teaching and assessment (Vol. 24). Springer International Publishing. https://doi.org/10.1007/978-3-319-66659-4
Stroh, D. P. (2015). *Systems thinking for social change: a practical guide to solving complex problems, avoiding unintended consequences, and achieving lasting results*. Chelsea Green Publishing.
Taylor, C. A. (2017). Is a posthumanist *Bildung* possible? Reclaiming the promise of *Bildung* for contemporary higher education. *Higher Education, 74*(3), 419–435. https://doi.org/10.1007/s10734-016-9994-y
Tonso, K. L. (2014). Engineering identity. In A. Johri & B. M. Olds (Eds.), *The Cambridge handbook of engineering education research* (pp. 267–282). Cambridge University Press.
United Nations. (2021). United Nations sustainable development goals. United Nations. https://sdgs.un.org/goals
Van Maanen, J. (1988). *Tales of the field: On writing ethnography*. University of Chicago Press.
Vanasupa, L., & Barabino, G. (2022). An engineering education of Holism: Einstein's imperative. In *Insights into global engineering education after the birth of industry 5.0* (pp. 77–100). InTech Open.
Vanasupa, L., Stolk, J., & Herter, R. J. (2009). The four-domain development diagram: A guide for holistic design of effective learning experiences for the twenty-first century engineer. *Journal of Engineering Education, 98*(1), 67–81. https://doi.org/10.1002/j.2168-9830.2009.tb01006.x
Varela, F. J. (1999). *Ethical know-how: Action, wisdom, and cognition*. Stanford University Press.
Vesilind, P. A. (2010). *Engineering peace and justice: The responsibility of engineers to society*. Springer.
Yancey, K. B. (2016). *A rhetoric of reflection* (1–1 online resource (x, 328 pages): illustrations (some color)). Utah State University Press. http://site.ebrary.com/id/11244346.
Yancey, K. B., & Rhodes, T. (2019). *EPortfolio as curriculum: Models and practices for developing students' ePortfolio literacy* (First edition, 1–1 online resource (xii, 274 pages)). Stylus Publishing LLC. https://search.ebscohost.com/login.aspx?direct=true&scope=site&db=nlebk&db=nlabk&AN=2097762

Sarah Appelhans, Assistant Professor of Engineering Studies at Lafayette College. She earned her PhD in Cultural Anthropology at the University at Albany (SUNY) where she conducted research on the cultural factors that contribute to inequalities in engineering. As a postdoc at Bucknell University, she was the resident ethnographer in the Electrical and Computer Engineering Department, exploring applications of Amartya Sen's capabilities approach in engineering education. Her current book project, *On the Bleeding Edge: Gender, Immigration and Precarity in Semiconductor Engineering*, investigates the intersections of gender, race/ethnicity, and immigration status among semiconductor engineers.

Alan Cheville, Professor of Electrical and Computer Engineering at Bucknell University, USA. He studied optoelectronics and ultrafast optics at Rice University, transitioned to faculty member at Oklahoma State University working on terahertz frequencies and engineering education. He left OSU to serve for two and a half years as a program director in engineering education at the National Science Foundation, then took a chair position in the Electrical and Computer Engineering Department at Bucknell University. He serves on several national-level advisory boards. Recently he helped found the Maker-E, an electronic MakerSpace for students, and is currently leading a department-wide effort to integrate more understanding of social sciences into the engineering curriculum.

Chapter 10
Usefulness and *Bildung*: Engaging Students in a *Bildung*-oriented Engineering Education

Ronny Kjelsberg and Magnus Strøm Kahrs

Abstract In the context of everyday higher education, the concept of 'usefulness' seems to be associated with the affordance of a certain skill or piece of information to be applicable/relevant in future professional contexts.

There has been a turn in engineering education toward seeing the future engineer as an agent in society, resolving perceivingly more complex problems, often in cooperation with representatives from other professions. This is reflected in recent policy documents on engineering education. This turn puts some pressure on our colloquial interpretation of 'usefulness', as it requires an expansion of skills and knowledge relevant to the students' future professional lives. In engineering education, the perceived future role of an engineer has changed in recent years, emphasizing the societal role of the engineer. As such, the scope of 'usefulness' as understood in an engineering context has expanded, including clear normative connotations beyond mere applicability. In the context of engineering education, this yields providing solutions that take into account ethical, aesthetical, economical, and sustainable aspects, as well as purely technical aspects.

Simultaneously, this more system-perspective and citizen-oriented turn in engineering education connects the education closer to the idea of *Bildung*, where students are expected not only to become practitioners of a craft, but also active participants in society.

This chapter discusses the contentious relationship between the concepts of *Bildung* and usefulness, where a theoretical discussion contrasting different traditions and forms of *Bildung* is illuminated by datasets from 1700 1st year bachelor engineering students across four consecutive years and these students' motivations toward *Bildung*-oriented topics.

The students' profession-oriented motivations and orientation toward usefulness suggest that a *Bildung*-oriented education based on the idea that there is a conflict

R. Kjelsberg (✉) · M. S. Kahrs
Department of Physics, NTNU-Norwegian University of Science and Technology, Trondheim, Norway
e-mail: ronny.kjelsberg@ntnu.no; magnus.s.kahrs@ntnu.no

A. Buch, S. H. Christensen (eds.), *Bildung for Engineering Education and Practice*, Philosophy of Engineering and Technology 49,
https://doi.org/10.1007/978-3-031-86581-7_10

between usefulness and *Bildung* will be unsuccessful in motivating these students to engage themselves in society. An education for engineering *Bildung* should thus be based on integrating the ideas of usefulness into the concept of *Bildung*.

Keywords Usefulness · *Bildung* · Engineering education · Social engagement

10.1 Intersections of *Bildung* and Usefulness

In the context of everyday higher education, the concept of 'usefulness' seems to be associated with the expectation of a certain skill or piece of information to be applicable or relevant in future professional contexts. As such, 'usefulness' is connected to the concept of relevance. Whether or not something is considered relevant is furthermore a matter of perspective: Relevance can be assessed at the individual level, at a cultural or disciplinary level, and at a societal level. If something were to pass as relevant through all these levels, there is limited need to motivate students, and vice versa—a concept, or a skill, that is considered to be relevant by society, but not by the students, would entail a more thorough motivation towards the students. In teaching *Bildung*-oriented topics to bachelor engineering students in Norway, faculty members have volunteered problems engaging students in topics like ethics, as the students don't see the topics as relevant for their perceived future work life (Thorvaldsen & Henne, 2017). We believe the negotiation between the concepts of "usefulness" and "*Bildung*" can shed light on this problem and point to paths toward students embracing a societal perspective on engineering.[1]

10.1.1 Usefulness

In this chapter we will use the term 'usefulness' rather than 'utility' as the latter is often used in a narrower sense in economic theory and philosophy than the wider concept we wish to discuss here, incorporating contribution to various individual and societal needs.

From a societal perspective, what is considered useful for an engineering student depends on the perceived roles engineers will have in society. The world our engineering students are part of and eventually will enter as professional agents could be described as a social world of complexity, in terms of problems, interests, pace of change, and consequences—a world that Elmose and Roth (2005) describe as a *risk society*. In this perspective, resolutions to problems are the results of collaboration among professionals from different disciplines—real-world problems are rarely resolved by representatives of one discipline, let alone individual representatives

[1] Some preliminary findings and analysis from this chapter were presented at the 2023 CDIO conference.

(Grasso et al., 2010). This description resonates with recent policy documents from the council Universities Norway on engineering education in Norway (UHR, 2020):

> Systems that interact make out an important feature of a modern society. You [the prospective engineer] must therefore be proficient in working independently, and in working in teams—with engineers from your own and other disciplines, experts from other professions, and in interdisciplinary teams. As an engineer, you are working with people, you are ethically responsible and environmentally conscious, and you have a great impact on society! (p. 5, our translation)

This perspective implies that the scope of what is considered useful or relevant for an engineer extends well beyond pure technical matters, which in turn calls for several engineering 'profiles', as described by Kamp and Klaassen (2016). Examples of this extension are ethical, aesthetical, economic, environmental, and sustainable considerations—considerations which are linked to a notion of *Bildung* for engineers.

The question we will pursue in this chapter is—how do societal perspectives on usefulness for engineers fit with engineering students' reported motives for pursuing an engineering education? Based on data from Norwegian bachelor engineering students, the answers to this question will yield recommendations about how *Bildung* for engineers for the twenty-first century could be implemented.

It is beyond the scope of this chapter to investigate the reasons behind engineering students' reported motivations for pursuing an engineering education. However, for illustrational purposes, we will take a quick look at how engineering education is framed by our institution. We will present how the bachelor program for construction engineering is framed, in 2019 and 2023. This is a somewhat speculative comparison, as we take a snapshot of what is to be considered manifestations of a continuous discussion about what it means to be an engineer. However, we believe that these excerpts illustrate the ongoing turn in engineering education.

Construction engineers design and calculate, construct, plan, and manage work together with others. In addition to buildings, a construction engineer works with roads and infrastructure, energy, water, and environment. As a student in the construction engineering program, you will be equipped to face today's challenges concerning construction, such as changes in environment and climate. You will also be fit to face the industry's expectations when it comes to innovation and productivity. Do you want to join in making good, sustainable solutions?	*Do you want to join in planning, building, roads, water, and sewerage, and at the same time provide for a greener future? The construction industry is the largest mainland industry in Norway and plays a central part in the adaptation towards a sustainable society. To obtain emission-free construction sites, houses that require little energy, and recycling of building materials, environmental considerations must enter into all construction plans. In addition, heavy precipitation, fluctuations in temperature, and an increased risk of floods, landslides, and storm surges, will create extra challenges for roads and constructions in Norway. At the construction engineering program, you will be equipped to face these challenges and the industry's expectations when it comes to innovation and productivity.*

What we can see from this is that the use of engineering-specific verbs (design, calculate, construct, plan, manage) was more pronounced in the 2019 version compared to the 2023 version. Environment and sustainability are mentioned in both versions. The main difference between the 2019 and the 2023 versions is perhaps that the engineer of 2023 is an agent set in a value-added context, solving important societal problems. This can also be seen in the changes in the study program portfolio. Engineering programs, such as construction engineering and mechanical engineering have a history going back roughly a hundred years. In recent years, however, these traditional engineering programs have been accompanied by new programs, for instance, engineering programs within renewable energy. This is in line with international developments where outside influences have encouraged engineering education to include wider societal perspectives (e.g. CDIO, 2022).

10.1.2 Bildung

The relationship between engineering and *Bildung* has long been contentious, and by some even seen as contrary. According to Sjöström et al. (2017), Wilhelm Von Humboldt maintained that language was the main manifestation of *Bildung*, and due to his influential position in academia, this led to the dismissal of the natural sciences' contribution to *Bildung*, for the benefit of the humanities. As an illustration of this dismissal of science's contribution to *Bildung*, Dietrich Schwanitz (1999) describes two high school sweethearts, Sabine and Torsten who plan to get married after their studies. Sabine goes to Hamburg to study psychology, German, and art history, while Torsten moves to Aachen to study mechanical engineering. While Sabine gains insight into the foundations of communication and the symbolic systems of culture and is transformed as a person, Torsten learns how to build machines. When they meet again after their studies, Sabine thinks Torsten sounds like a Neanderthal and breaks off the engagement (Schwanitz, 1999). Here, the practical knowledge of the engineer is presented as the antithesis of *Bildung*. This view is rooted in traditions of *Bildung* where notions of usefulness are seen as conflicting with *Bildung*. Other traditions, however, see it differently, and within more recent reforms of engineering education in several countries, *Bildung*-oriented views have been very present, as illustrated by the quote from the policy document from Universities Norway above.

Although it seems fairly easy to find accounts of what does *not* count in different interpretations of *Bildung*, it is a concept that can be hard to define concisely. However, many descriptions connect the term to education making not only professionals who are skilled at their craft but also active and engaged citizens (Adler, 1952; Klafki, 2016), contributing with their professional expertise for the betterment of society. This account of *Bildung* fits well with the perception of the role of our future engineers. *Bildung* and an extended notion of usefulness seem to coincide in this perspective. However, this relationship has historically been contentious, as we will discuss below.

Building on previous work by Gustavsson (Gustavsson, 2012, 2014) and Burman (2011) Sjöström et al. (2017) describes five traditions of *Bildung*: (1) a *classical*

Bildung based on the ideas by Willhelm Von Humboldt, which is described as a process where the individual develops through education, (2) an Anglo-American tradition of *liberal education*, where humanism and generalization are central values, often associated with a classical canon, (3) the less academically oriented Scandinavian *folk-Bildung*, which focuses on a *Bildung* for all citizens, not just for the formally educated, (4) a *democratic education*, which also focuses on the collective aspect, albeit towards formal schooling, and (5) a *critical-hermeneutic Bildung* or *Allgemeinbildung*, which share similarities with folk-*Bildung*, i.e., *Bildung* for all citizens, but with an explicit emphasis on *Bildung* in all human capacities.

One of the first things to notice from this initial summary is a progression from emphasizing *Bildung* as an individual process of emancipation, to seeing *Bildung* as a collective process both of emancipation, but also toward the betterment of society. The latter is in line with current ideals within Norwegian engineering education, calling for an engineering education that e.g. "facilitates the interaction between ethics, environment, technology, individual and society" (UHR, 2020, p. 28), and with the expected development of the future of engineering education internationally (Graham, 2018, p. 43).

In addition, we can separate ideas about *Bildung* in different forms. Wolfgang Klafki, in an influential paper from 1959 (updated in 1979), describes a difference between *material* and *formal* theories of *Bildung* (Klafki, 2001). The material theories deal with the contents of *Bildung*, i.e., knowledge and skills. In contrast, the *formal* theories concentrate on the person that is going through a process of *Bildung*. These two theories can also be connected to the contemporary educational philosophies of essentialism and progressivism, where the former, like material theories of *Bildung*, are curriculum-oriented, while the latter concentrates on the formation of the student, similar to the formal theories of *Bildung*. Up against both these two theories, Klafki sets his concept of *categorical Bildung*, in which the individual and the world simultaneously open themselves up to one another (Klafki, 2001, p. 193). Sjöström et al. (2017) on the other hand complements the two philosophies with reconstructionism, shifting the focus of a *Bildung*-oriented education toward the effects a knowledgeable student can have on society through critical citizenship. We can thus also discern two different perspectives on *Bildung* here, where we in addition to the traditional view of *Bildung* as a *process*, also can view it as a *product*, seen as an effect on society. The critical citizenship-perspective is also in line with a perspective from Jon Hellesnes, describing *Bildung* as the process of connecting the professional language to the everyday language, specifically the language of politics (Hellesnes, 1992).

10.1.3 **Bildung** *and Usefulness in Engineering Education*

As described by Sjöström and Eilks (2018), Classical *Bildung* is primarily concerned with the emancipation of the individual, and can thus be seen as formal *Bildung*, following Klafki's categorization (Sjöström & Eilks, 2020). The democratic education tradition is more focused on the individual's development and is as such to be considered formal. The same can be argued about the English liberal education tradition (Sjöström & Eilks, 2020). The American "Great Books"-tradition which is also part of the *liberal education* tradition can in contrast be seen as a *material* theory of *Bildung*. This tradition is deeply connected to the idea of a *canon*—a list of *great books* every educated person should be familiar with (Liedman, 2001) which stands in contrast to vocational competence (Sjöström & Eilks, 2020).

To the Scandinavian folk-bilders on the other hand, there was no conflict between *Bildung* and practical vocational knowledge. It was exactly the practical applications that made the natural sciences a popular arena for folk-*Bildung*. A Norwegian edition of the Swedish "Textbook in the Natural Sciences for Elementary Schools and Elementary School Teachers Seminars" by Nils Johan Berlin sold over 40,000 copies over the first 4 years from 1853 onwards. In comparison, the first complete Norwegian Bible translation from 1854 had a first printing of 6000 copies (Roos, 2017). The reason for this popularity was its usefulness and relevance among a general population largely living off and from nature. That the usefulness from the technical and natural sciences formed a gateway and a motivation toward *Bildung*, and learning to read was more important to the folk-bilders than exactly what is read. This points to this tradition being a more *formal* tradition of *Bildung*, using Klafki's terminology. However, one could argue that for the reading peasants, what was being read was of paramount importance, as its usefulness was central to their motivation. The categorization as formal *Bildung* is thus based on a top-down perspective seen from the perspective of Grundtvig, Ole Vig, etc. A bottom up-perspective seen from a Scandinavian peasant would perhaps categorize it as categorial and as such more similar to the critical-hermeneutic *Bildung* (Sjöström & Eilks, 2020) than a traditional perspective would allow.

The current discourses around engineering education are about expanding the perspective of engineers into the societal context, connected to topics such as sustainability, cross-disciplinarity, and diversity (Borrego & Newswander, 2008; Smith-Doerr et al., 2017; Wan Alwi et al., 2014). This is often implicitly discussed in relation to the concept of *Bildung*. However, we believe that these discussions could be informed by the long and multi-faceted history of scholarship on the concept of *Bildung*.

There have been several attempts to include *Bildung* in engineering education. One such example comes from Axelsson (2009) who describes the historical development of *Bildung*-oriented topics in engineering education at Chalmers University, Sweden. Chalmers started by teaching technical subjects with a societal connection (e.g. "Technical knowledge - increased security"), later moved in a more general humanistic *Bildung* direction, and finally moved toward a system perspective of engineering (Axelsson, 2009, p. 230). The initial development points to two competing perspectives: Should the *Bildung* aspects of engineering education lead to a general humanistic-social science-oriented *Bildung*, or should they help to give engineers a system perspective on technology? (Axelsson, 2009). This discussion partially mirrors the discussions of the "great books" and the "Scandinavian folk-*Bildung*" traditions, as the more general *Bildung* perspective does not imply usefulness for the engineer, while the system perspective does.

We can thus see a duality in the view on the relation between *Bildung* and usefulness not only in theories on *Bildung* but also in the concrete attempts to create a *Bildung*-oriented engineering education.

10.2 Data Collection and Analysis

The data in this study has been collected from first-year bachelor engineering students at the Norwegian University of Science and Technology, through an open question on why they have chosen their field of study. It has been collected at the start of the course "Introduction to the engineering profession" from 2019 to 2022, about a month into the students' first semester.

The data consists of open-text responses that vary in length from single-word responses to a few short sentences, and these have been categorized using thematic analysis (Braun & Clarke, 2006). The open-text responses were coded thematically, and the relative frequency of the different themes was then registered quantitatively.

All responses have been collected by the online student response systems iLike (2019–2021) and Mentimeter (2022), in a combination of online or live classes (2019) or online classes (2020–2022). Note that the wording of the question was slightly changed from 2019 to 2020 (to compare with another student group) from "Why do you want to become an engineer?" to "Why do you want to study engineering?". This must be kept in mind when analyzing the data, and in coding it has affected the interpretation of e.g. single-word responses like "interesting" or "cool" when it comes to connecting these to interest to "subject" or to "job". The brevity of each data point also led us to chunk responses into relatively large categories, which contain both explicit and implicit responses which we inferred to pertain to a certain category. This process is described more in detail in the Results section. The wording of the questions did not seem to have any notable effects when coding other themes. Upon examining the data we have decided to include the 2019 data as well as the results align well with those from other years.

The scope of the project has been slightly expanded over the period. From 2020 the students were in addition asked about their background (gender, parents' education, minority language), and in 2022 they were in addition asked about their study program.

The coding of themes has been done by an initial coding by the first author, an evaluation and re-coding of parts of the data by the second author, a discussion around different choices of coding, and a subsequent re-coding of the entire material based on agreed choices.

10.3 Motivations of Engineering-Students

The student's responses disclose a strong connection both to their future profession and to different aspects of usefulness, as disclosed in the themes discovered in the analysis.

10.3.1 Themes

We have separated the category of intrinsic motivation into a category for interest in and enjoyment from the subjects the students are studying, and another for interest in and (perceived) enjoyment from the job/the profession they envision themselves moving into after their education. These two can however sometimes be hard to discern when students simply respond with "interest", "exiting" etc. without explicitly connecting these motivations to either. In addition, we have the category "Altruistic" which describes different motivations for contributing to society (helping with the climate crisis, developing technology to improve society, etc.).

In categorizing Extrinsic motivations, we have concentrated on responses related to salary and status.

We have separated another category—"possibilities job" to encompass motivations connected to job safety, or "many possibilities", but most of all to the idea of a varied future workday and the more overarching idea of a multitude of work possibilities. In many responses, this is connected to the lack of this in a current job after vocational education.

We have also made a combination category summing up the proportion of students who have responded to at least one category explicitly connected to their future job as an engineer. These are both the "interest job", "altruistic", "extrinsic" and "possibilities job" categories.

The category of "influence" encompasses references to influence from parents or other relatives, teachers, or even celebrities when it comes to choosing a future education and profession.

In categorizing "Previous education", all references to previous education as motivation have been included like experiences of mastering these topics in upper secondary education or viewing engineering as a natural next step after a vocational education, including after a period of employment.

The theme "understanding" mainly consists of a wish to understand (more about) how different technologies or technological objects work.

In categorizing "Challenge" both explicit references to wanting to "challenge" oneself are included, but also some more general comments on wishing to develop oneself that allude to challenging.

In categorizing Creating, both references to "creating", "developing", "building", "forming" and some slightly more general references to creativity and shaping are included.

There naturally is some overlap between this category and the "practical", but both the more general references and the development of e.g. computer programs that are not considered as "practical" in a more physical sense, create a difference. On the other hand, in the "practical" theme some references to the education and/or profession as more practically oriented will not be included in the "creating" category.

In categorizing usefulness, both explicit references to "useful" and "usefulness" and e.g. descriptions of "problem-solving" and similar descriptions pointing towards something being useful have been included.

In the theme "future" references to their education being future-oriented is notable, but other references to the future are also included, including uses of "new" (e.g. creating something new, previously unseen) and similar terms pointing to the future.

In categorizing the theme "practical", both explicit references to practical work and references to creating/developing concrete "things" have been included.

In addition, we register all students with responses within the four last themes to a combined "C,U,F,P" construct, from an interpretation that these responses are often connected to a certain view of engineering, e.g., creating something practical for the future is useful. This is a relatively common theme in the responses, we can also see that of a total of 599 students who in brief responses have been assigned one of these themes, 250 have had at least one of the others assigned as well, and the overrepresentation of the other categories among the students motivated by usefulness shows the same (Table 10.1).

Examples of students expressing a combination of these motivations can still be brief and compact: «to find solutions to practical and future problems», "Creating something new and contributing to technological development" etc.

Finally, a proportion of students answered the questions with obvious jokes. These have been the hardest to categorize. While keeping in mind the well-known aphorism from Piet Hein: "Taking fun as simply fun and earnestness in earnest shows how thoroughly thou none of the two discernest.", we have not dismissed responses that would otherwise have been categorized, simply because they have been made in a joking tone, e.g. a response of "$$$" will have been categorized as extrinsic in the same way as "good salary". On the other hand, we have had a lower threshold for dismissing these from categories other than the "joke" category as it is harder to know what the student means. The number of respondents with joking answers is however sufficiently low for it not to have any large disturbing effect on the overall data.

To simplify the reading of the data, we have limited the presentation of the data in Table 10.1 to the parts most central to the discussion, while all themes and categories of students are presented in an appendix directly after the chapter (Appendix 1).

10.3.2 Student's Motivations

Overall, we can see from Table 10.1, engineering students have a strong interest in both the subject of engineering, but most of all their future profession as engineers. 70% of students volunteer a motivation that is in one way or another related to their perceived future profession or work life. There is also a strong motivation connected to the theme combining practical, useful, creative, and future-oriented elements. The relevance of this construct can also be seen by how the students categorized in the "useful" theme, are overrepresented in all the other themes in the construct.

There is also a strong altruistic motivation among engineering students, which however has declined over the 4 years. In addition, we see several less prominent,

Table 10.1 Frequency of the most central coded themes and sets of students for this paper, from the answer to "Why do you want to study engineering?", for different groups of students. For the full set of themes and examined sets of students, see Appendix 1. The frequencies show what proportion of respondents gave responses in line with the different themes. Rows represent coded themes. "Job combined" represents all students with at least one of the various job-related themes, and "C+U+F+P similarly lumps students with Create, Useful, Future, and Practical themes. The columns represent different (sub)groups of students. 2019-2022 represent the different years, Female and male represent all students who self-categorized as such from the years we have this data (2020–22). Useful represents all students (2019–2022) whose response was themed in the "useful" theme. Total represents all students

Groups Themes	2019 $N = 427$	2020 $N = 490$	2021 $N = 399$	2022 $N = 384$	Female 2020–22 $N = 290$	Male 2020–22 $N = 878$	Useful 2019–22 $N = 156$	Total 2019–22 $N = 1700$
Interest subject	.2505	.5122	.4286	.2995	.4450	.3796	.2821	.3794
Interest altruistic	.2763	.1877	.1228	.0885	.1498	.1148	.4295	.1724
Job-combined	.7517	.6490	.6867	.6875	.5950	.6516	.7628	.6906
Previous ed.	.0632	.1143	.0952	.0755	.0489	.1045	.0320	.0888
Understanding	.0257	.0082	.0351	.0208	.0073	.0272	.0385	.0218
Challenge	.0889	.0571	.0576	.0469	.0456	.0522	.0641	.0629
Create	.3255	.1918	.2306	.1406	.1410	.1942	.6218	.2235
Useful	.1170	.1122	.0777	.0547	.0870	.0818	1	.0924
Future	.1545	.1163	.1454	.0964	.1511	.1066	.2244	.1282
Practical	.0843	.1306	.0727	.0599	.0904	.0920	.1731	.0894
C,U,F,P-comb	.0843	.1306	.0727	.0599	.3161	.3110	1	.3535

but recurring forms of motivation registered in Table 10.1, as described in more detail above.

When it comes to results that are particularly relevant to the topic of this chapter there are several observations we should make from the data.

We see that both the altruistic and the combined "CUFP" construct decline somewhat over time.

Regarding our topic of usefulness, we can see a strong correlation between the group of students categorized in the "useful" theme and altruism, with a 43% altruistic motivation vs 17 & for students as a whole. We can thus see a strong and very significant overrepresentation of altruism among the students who are motivated by usefulness. Doing a chi-squared test comparing these students to all students yields a chi-square value of 58.79 with a corresponding p-value of 1.75×10^{-14}, meaning the null-hypothesis can be rejected and we can conclude a strong statistically significant correlation between being a student motivated by usefulness and being motivated by altruism.

Although there is a strong and very significant overrepresentation of altruism among the students who are motivated by usefulness, there are some results for subgroups that might suggest reservations. Notably renewable energy students are not surprisingly overrepresented in altruism, but they have a very low score on "useful" (See Appendix 1 for data on study programs). This could suggest that there is a conflict between the view on usefulness and altruism for some student groups. The low N (34) however should make us weary of drawing to strong conclusions. Conducting a chi-squared test comparing renewable students to all 2022 students on the ratio of altruism vs. usefulness, nevertheless, yields a chi-square value of 4.56 and a p-value of 0.033, which is statistically significant.

We can also see a slight overrepresentation of females on altruism, without seeing the same on usefulness, but doing a similar test on female vs. male students yields a chi-square value of 0.298 and a p-value of 0.585, making the difference not significant.

10.4 Usefulness, *Bildung*, and Motivation: Discussion and Recommendations

There are reasons to be wary of making strong conclusions on development over time in the student group. One element is the slight change in the wording of the question from 2019 to 2020. Another is the change in setting from a mix of online and physical lectures to only online lectures for collecting data. Although the form of response was identical, the different user interfaces and experiences with the response system might still have subtle influences on the students' responses. Only in 2020 and 2021 have the data thus been collected in a completely identical manner. These years also include both the rise and decline of a Covid-pandemic, economic uncertainty and the start of the Russo-Ukrainian war. For these reasons, we will mainly discuss the cumulative results from these 4 years, where the data are more robust.

Keeping in mind these uncertainties, we can nevertheless briefly comment on a couple of developments over time from the results (see Table 10.1 for data). One is the decline in altruistic motivations. Could the insecurities from the pandemic and war with consequential economic insecurities have shifted students' focus away from e.g. tackling climate change and more toward securing their own future? The extrinsic motivation, however, seems stable. There is also a decline over time in the "create, develop, build" theme. Are students becoming less practically and more academically oriented? Our data cannot yield conclusions on this, but these could be topics relevant for further, more qualitative, studies. In the following, we will discuss the accumulated results from all years.

When discussing *Bildung*, the category of altruistic motivation is of particular importance as it entails contributing to society and as such positions the engineer as a conscious actor in society, in line with the definitions of *Bildung* we have discussed (Hellesnes, 1992).

Regarding our main topic of usefulness, we see that there is a strong overall motivation among engineering students toward the practical, concrete, and useful, including contributing to society both in concrete areas like climate change, but also more generally in contributing to technological development. The idea of something being useful here clearly has a broader impact than the concrete references to usefulness included in the "useful" theme itself. Building, creating, and contributing to technological development is a central motivation for engineering students, and the considerable overrepresentation of usefulness-motivated students in both the other elements in the combined "CUFP"-construct confirms this. This could be used as leverage for integrating *Bildung* perspectives in engineering education.

The spectrum of expressed motivations among the students does, however, put some restraints on how this can be integrated.

The relatively recent and quick shift into environmental concerns and sustainability in engineering education may have as a consequence that you have a mix of students, some attracted by the "new" profile, while others by the traditional profile of engineering. In the data, this can e.g. be reflected by where renewable energy students differentiate themselves from other student groups with more focus on altruism and less on usefulness. This is however not reflected in the average student, where usefulness has a very positive correlation with altruism. The latter can be connected with a more traditional view of the engineer as a "builder of society" particularly prominent in industrialization processes in the previous century (Nygaard, 2013).

10.4.1 Positioning Engineering **Bildung** *Within a Theoretical Framework*

As the data shows engineering students' motivations are focused on their future work as an engineer. This is not surprising, considering the context of 3-year 180 ECTS bachelor professional bachelor education, and this work focus includes their altruistic motivation. Among the themes, altruism has the strongest connection to

Bildung, in the context of connecting their profession to society. In line with this, the perspective on *Bildung* will be more in line with *Bildung* as a product or result of the students' development process rather than the process itself, mirroring one half of this dichotomy described by e.g. Fischler (2011).

In utilizing a definition of *Bildung* that connects students and their profession to society, we can see that the elements we have to build upon from the students' motivations go from the altruistic "contribute to fighting the climate crisis through developing renewable energy" to the more purely technology-oriented "contributing to developing new technology" to the more generic "problem solving" and to the wish to develop concrete new products and solutions. It should be noted that expressed motivations like "problem-solving" might not bear any reference to society, but rather reflect an intrinsic interest in solving problems. Nevertheless, as resolutions to technological problems are inevitably linked to society, we can utilize the educational context of engineering education to include both intrinsic interest and societal usefulness, thus supporting a systems perspective on *Bildung* in which usefulness is an integral part. This is supported by the overrepresentation of students motivated by usefulness *and* altruistic motivations.

We can thus discern traces of a taxonomy of usefulness in students' motivations, from the very concrete useful product, more connected to the colloquial "usefulness" to the more overarching societal perspective which can connect well with a wish to create a *Bildung*-oriented engineering education. Usefulness, in one form or another, is thus at the core of engineering students' societal motivations, and an interpretation of *Bildung* that separates *Bildung* from usefulness will be of little relevance to engineering students' existing motivations. One could envision an education attempting to move engineering students upwards in their taxonomy of usefulness, from building things to contributing to technological development in general toward building and improving society.

This would entail building more on the traditions like the Scandinavian "folk-*Bildung*", and less on material "great books"-traditions. It also supports the more integrated approach of including societal, ecological, etc. perspectives in engineering courses rather than separating them out in their own courses. This is also in line with current trends in engineering education, promoting contextualization of the curriculum (FTS, 2020).

We can see one distinct similarity with the Scandinavian folk-*Bildung* tradition. While the folk-Bilders were more interested in learning peasants to read than in what they read, thus positioning this as a formal theory of *Bildung*, the peasants, like the engineering students, were not neutral to the content. Reading about nature was useful for them, and as such they chose this, making the *Bildung* process have material elements, but in a bottom-up sense.

In addition to the discussed traditions and forms of *Bildung* and viewing *Bildung* as a process vs. a product, we can thus add top-down vs. bottom-up as an additional dimension when attempting to understand the complexity of the concept of *Bildung*. This will be particularly important for traditions like the critical hermeneutic who promote engaging with the reality and the perspective of the student (Fischler, 2011). This is in line with Klafki's ideas on *Bildung*, promoting making science meaningful for the student and lifting it into the student's horizon of understanding

(Hohr, 2011; Klafki, 2001). This makes its results more similar to a categorial and critical hermeneutic *Bildung* than seen at first glance. In forming a *Bildung*-oriented engineering education we must attempt to make this connection in a conscious manner, integrating the engineering students' professional and usefulness-oriented motivations with the altruistic motivations, to make a connection to society and as such give students a wider critical perspective on their role in society, preferably thus moving closer to a critical-hermeneutic and categorial *Bildung*.

10.4.2 Consequences for Engineering Education

For a *Bildung*-oriented engineering education to be successful, it must thus dip into engineering students' existing motivations that are centered around different forms of usefulness/relevance. This would be in line with the system perspective described at Chalmers (Axelsson, 2009). Such a perspective, being more overarching, can integrate the general *Bildung* perspectives with the professional usefulness in a better sense than the more limited technical subjects with societal connection.

Creating a curriculum that at the same time seems useful to students and thus motivates them and makes them open up to the ideas in this curriculum, which then again transforms them and changes them as human beings. As discussed above, this parallels the processes from the Scandinavian Folk-*Bildung* tradition but also points toward something similar to the critical hermeneutic tradition and to Klafki's idea of *categorial Bildung* (Klafki, 2001). Based on our data, integrating the idea of usefulness would be a key to success in *Bildung*-oriented engineering education.

Appendix 1: Full Set of Coded Themes on the Full Set of Categories of Students

Frequency of all the coded themes described in subchapter 15.3.1 from the answer to "Why do you want to study engineering?" over different years. The frequencies show what proportion of respondents gave responses in line with the different themes. Rows represent coded themes. "Job combined" represents all students with at least one of the previous job-related themes, and "C,U,F,P-comb" similarly lumps students with Create, Useful, Future, and Practical themes. The column represents different (sub)groups of students. 2019–2020 represents all students from these different years. Female and male represent all students who self-categorized as such from the years we have this data (2020–22). The Construct- Mechanic columns represent students divided into study programs for 2022 (the only year such data was collected). Useful represents all students (2019–2022) whose response was themed in the "useful" theme. Total represents all students.

Groups Themes	2019 $N = 427$	2020 $N = 490$	2021 $N = 399$	2022 $N = 384$	Female 2020–22 $N = 290$	Male 2020–22 $N = 878$	Construct 2022 $N = 64$	Comput 2022 $N = 66$	Electric 2022 $N = 61$	Renewab 2022 $N = 34$	Mechanic 2022 $N = 62$	Useful 2019–22 $N = 156$	Total $N = 1700$
Interest subject	.2505	.5122	.4286	.2995	.4450	.3796	.3281	.3030	.3278	.2058	.2580	.2821	.3794
Interest job	.3957	.3204	.5088	.4766	.3517	.4327	.4219	.4696	.5573	.5294	.5806	.4808	.4212
Interest_altruistic.	.2763	.1877	.1228	.0885	.1498	.1148	.0469	.0606	.0983	.2058	.0967	.4295	.1724
Extrinsic	.1943	.1265	.1303	.2135	.0904	.1616	.1875	.3030	.2295	.1764	.1935	.0769	.1641
Possibilities job	.2084	.2163	.1679	.2708	.2443	.2013	.1562	.3787	.2622	.2058	.3387	.1026	.2129
Job-combined	.7517	.6490	.6867	.6875	.5950	.6516	.6094	.7575	.7213	.7647	.7903	.7628	.6906
Influence	.0163	.0408	.0351	.0313	.0423	.0272	.0938	0	0	0	.0322	0	.0312
Previous ed.	.0632	.1143	.0952	.0755	.0489	.1045	.0781	.0303	.1967	.0882	.0967	.0320	.0888
Understanding	.0257	.0082	.0351	.0208	.0073	.0272	.0156	0	.0327	.0294	.0161	.0385	.0218
Challenge	.0889	.0571	.0576	.0469	.0456	.0522	.0469	.0151	.0491	.1176	.0161	.0641	.0629
Create	.3255	.1918	.2306	.1406	.1410	.1942	.125	.1666	.1475	.0588	.3064	.6218	.2235
Useful	.1170	.1122	.0777	.0547	.0870	.0818	.0156	.1060	.0327	.0294	.0806	1	.0924
Future	.1545	.1163	.1454	.0964	.1511	.1066	.0313	.1515	.1147	.1176	.1451	.2244	.1282
Practical	.0843	.1306	.0727	.0599	.0904	.0920	.0469	.0454	.0327	.0588	.1774	.1731	.0894
C,U,F,P-comb	.4426	.3612	.3609	.2396	.3161	.3110	.1875	.2727	.2295	.2647	.4354	1	.3535
Joke	.0374	.0204	.0350	.0677	.0103	.0406	.0469	.0606	.0491	0	.0322	.0064	.0388

References

Adler, M. J. (1952). *The great ideas: A syntopicon of great books of the western world* (Vol. 1). Encyclopedia Britannica.

Axelsson, A.-S. (2009). Bildning i den tekniska högskolan [Bildung at the engineering college]. In Å. Ingerman, K. Wagner & A.-S. Axelsson (Eds.), *På spaning efter teknisk bildning [Searching for technical Bildung]* (pp. 222–241). Liber.

Borrego, M., & Newswander, L. K. (2008). Characteristics of successful cross-disciplinary engineering education collaborations. *Journal of Engineering Education, 97*(2), 123–134. https://doi.org/10.1002/j.2168-9830.2008.tb00962.x

Braun, V., & Clarke, V. (2006). Using thematic analysis in psychology. *Qualitative Research in Psychology, 3*(2), 77–101. https://doi.org/10.1191/1478088706qp063oa

Burman, A. (2011). Svar på frågan: Vad är medborgerlig bildning? [Answer on the question: What is civic *Bildung*?]. In A. Burman (Ed.), *Våga veta! Om bildningens möjligheter i massutbildningens tidevarv [Dare to know! About the opportunities for Bildung in a time of mass education]* (pp. 9–31). Södertörns högskola.

CDIO. (2022). Optional Standard 1: Sustainable development. In *CDIO Optional Standards 3.0*: CDIO.

Elmose, S., & Roth, W. M. (2005). Allgemeinbildung: Readiness for living in risk society. *Journal of Curriculum Studies, 37*(1), 11–34. https://doi.org/10.1080/0022027041000229413

Fischler, H. (2011). Didaktik – An appropriate framework for the professional work of science teachers? In D. Corrigan, J. Dillon, & R. Gunstone (Eds.), *The professional knowledge base of science teaching* (pp. 31–50). Springer Netherlands.

FTS. (2020). *Bærekraftig kompetanse Delrapport 1 Fremtidens teknologistudier*. NTNU.

Graham, R. (2018). *The global state of the art in engineering education*. Massachusetts Institute of Technology.

Grasso, D., Burkins, M. B., Helble, J. J., & Martinelli, D. (2010). Dispelling the myths of holistic engineering. In D. Grasso & M. B. Burkins (Eds.), *Holistic engineering education*. Springer.

Gustavsson, B. (2012). Bildningens traditioner i transformation [The traditions of *Bildung* in transformation]. In A. Burman & P. Sundgren (Eds.), *Svenska bildningstraditioner [Swedish Bildung traditions]* (pp. 309–327). Daidalos.

Gustavsson, B. (2014). *Bildung* and the road from a classical into a global and postcolonial concept. *Confero, 2*, 109–131.

Hellesnes, J. (1992). Ein utdana mann og eit dana menneske. In E. L. Dale (Ed.), *Pedagogisk filosofi* (pp. 79–104). Ad Notam Gyldendal.

Hohr, H. (2011). Kategorial danning og kritisk-konstruktiv didaktikk. Den didaktiske tilnærmingen hos Wolfgang Klafki. In K. S. S. Dobson (Ed.), *Danning. Introduksjon til et ullent pedagogisk landskap* (pp. 163–175). Tapir Akademiske Forlag.

Kamp, A., & Klaassen, R. G. (2016). Impact of global forces and empowering situations on engineering education in 2030. Paper presented at the *12th international CDIO conference*.

Klafki, W. (2001). Kategorial dannelse: Bidrag til en dannelsesteoretisk fortolkning av moderne didaktikk. In E. L. Dahle (Ed.), *Om utdanning: Klassiske tekster*. Gyldendal Norsk Forlag.

Klafki, W. (2016). *Dannelsesteori og didaktik - nye studier*. Forlaget Klim.

Liedman, S.-E. (2001). *Ett oändligt äventyr – Om människans kunskaper*. Albert Bonniers Förlag.

Nygaard, P. (2013). *Ingeniørenes Gullalder - De norske ingeniørenes historie*. Dreyer.

Roos, M. (2017). Naturvitenskap som dannelsesprosjekt. In M. R. J. Tønnesson (Ed.), *Sann opplysning? Naturvitenskap i nordiske offentligheter gjennom fire århundrer*. Cappelen Damm Akademisk.

Schwanitz, D. (1999). *Bildung: Alles was man wissen muß*. Eichborn Verlag.

Sjöström, J., & Eilks, I. (2018). Reconsidering different visions of scientific literacy and science education based on the concept of *Bildung*. In Y. J. Dori, Z. R. Mevarech, & D. R. Baker (Eds.), *Cognition, metacognition, and culture in STEM education: Learning, teaching and assessment* (pp. 65–88). Springer International Publishing.

Sjöström, J., & Eilks, I. (2020). The *Bildung* theory—From von Humboldt to Klafki and beyond. In B. Akpan & T. J. Kennedy (Eds.), *Science education in theory and practice: An introductory guide to learning theory* (pp. 55–67). Springer International Publishing.

Sjöström, J., Frerichs, N., Zuin, V. G., & Eilks, I. (2017). Use of the concept of *Bildung* in the international science education literature, its potential, and implications for teaching and learning. *Studies in Science Education, 53*(2), 165–192. https://doi.org/10.1080/03057267.2017.1384649

Smith-Doerr, L., Alegria, S. N., & Sacco, T. (2017). How diversity matters in the US Science and engineering workforce: A critical review considering integration in teams, fields, and organizational contexts. *Engaging Science, Technology, and Society, 3*, 123–134.

Thorvaldsen, P., & Henne, I. (2017). Irrelevant! Møte mellom to kulturer. Paper presented at *the MNT-konferansen 2017*, Oslo.

UHR. (2020). *Nasjonale retningslinjer for ingeniørutdanning [National guidelines for engineering education]*. Universitets- og høgskolerådet. Retrieved from https://www.uhr.no/_f/p1/i3f80bc0d-b1b6-4334-bf25-7571488e9612/nasjonale-retningslinjer-for-ingeniorutdanning-vilkar-for-siving-vedtatt-nov-2022.pdf

Wan Alwi, S. R., Manan, Z. A., Klemeš, J. J., & Huisingh, D. (2014). Sustainability engineering for the future. *Journal of Cleaner Production, 71*, 1–10. https://doi.org/10.1016/j.jclepro.2014.03.013

Ronny Kjelsberg, Assistant Professor of Physics at the Department of Physics at NTNU, presently deputy head of department for education. He holds an MSc in theoretical physics and has worked in engineering education since 2003. He has been responsible for developing the course "Introduction to the engineering profession" since 2011, attempting to integrate Bildung elements in engineering education, and has written the textbook for the course which came with its 2nd edition in 2022. His research and development interests include the systemic interplay between science, technology end society and how to integrate the knowledge of this in STEM education.

Magnus Strøm Kahrs, Associate Professor in Science Education at the Department of Physics at NTNU. He earned his PhD in Science Education from NTNU in 2014 and has since been involved in the development of teaching science towards engineering students. Since 2018, Kahrs has been part of the University Science Education Research group at NTNU. Between 2019 and 2021 he served as Vice Dean for professional studies at the Faculty of Natural Sciences. His research interests include collaborative learning and educational development.

Part III
Bildung in Engineering Across Cultural Contexts

Chapter 11
Bildung and Education for Engineering Practice

James P. Trevelyan

Abstract *Bildung* is commonly represented in English with terms such as 'character building' or 'character formation'. Advocates have argued that school education should prepare young people to be active citizens in society and capable of moral judgements, that education should be more than a utilitarian preparation for employment. This chapter argues that widespread and costly engineering failures are associated with the lack of serious engagement with socio-technical aspects of engineering in undergraduate education. Research has demonstrated that engineers practice their technical expertise within complex social and organizational environments for which few education programs provide significant preparation. The lack of industry experience and appreciation of socio-technical factors among engineering faculty staff impedes attempts to bridge this gap. In the meantime, it is possible to learn from observations on engineering practice related to the consequences of an education system focusing primarily on the technical knowledge required by engineers. We can see how education practices shape the expectations and habits of engineers and constrain their ability to perform in practice. The chapter suggests how education restructuring might be approached in a slow process of reform.

Keywords *Bildung* · Engineering education · Engineering practice · Global South · Engineering failures

---

J. P. Trevelyan (✉)
The University of Western Australia, Western Australia, Australia
e-mail: james.trevelyan@uwa.edu.au

A. Buch, S. H. Christensen (eds.), *Bildung for Engineering Education and Practice*, Philosophy of Engineering and Technology 49,
https://doi.org/10.1007/978-3-031-86581-7_11

11.1 Introduction: Philosophical Ideas of *Bildung*

In this chapter, I argue that professional engineering education could be significantly enriched by helping students to become competent social actors by understanding their own personal development framed by ideas derived from the German concept of *Bildung*.

Bildung is usually represented in English with terms such as 'character building', 'character formation' or even 'well-rounded education'. Advocates, particularly in Germany and Scandinavia over nearly two centuries, have argued that school education should prepare young people to be active citizens in society and capable of moral judgements, that education should be more than a utilitarian preparation for employment. John Dewey, Martha Nussbaum and Amartya Sen have advanced similar arguments more recently (Kovalainen, 2012, pp185–186; Nussbaum, 2003, 2009; Williams, 2003), proposing that education should be for human development and freedom, not just for profit-making through corporate employment. Nussbaum, specifically, argued for attributes to be developed such as critical thinking, knowledge of the history and character of the heterogeneous groups that make up human societies, and the ability to see the world through the eyes and ideas of other people. In doing so, she echoed Wilhelm von Humboldt's *Bildung* vision from two centuries earlier based on reading, writing, arithmetic at the elementary level, followed by foreign languages, history, mathematics (Konrad, 2012, p 116). At the time, observing the chaos following the French revolution in the 1790s, he wrote "Man needs *Bildung* in order to find his way in the modern world, a world in a state of upheaval which no longer offers any kind of stability."

Bildung challenges the naïve reader, particularly in English, because it has been described in so many ways by different philosophers and advocates for its application in education. As Max Horkheimer warned us 70 years ago:

> Don't expect me to define it (*Bildung*). There are areas in which clear and simple definitions are more than to the purpose, and the role of definitions in knowledge should not be underestimated in any way. If one nonetheless seeks to grasp that which is relevant and substantive, which expresses itself in an idea, one must attempt to gain entry to an idea's internal life, its tensions and ambiguities, and surrender to the risk of colliding with contradictions in such a manner that one is even guilty oneself of contradiction. Suspect the need for intellectual purity of an exaggerated type that always requires in a discussion precise advance knowledge of what an idea means before it can be used at all. (Horkheimer, 1953), cited in (Siljander & Sutinen, 2012)

In a general sense, drawing on the ideas of Johann Amos Comenius (1592–1670) who wrote about human development in the image '*bild*' of God (Konrad, 2012, p. 108), Eetu Pikkarainen (2012) has posited that *Bildung* refers to the interaction between the self and the world and to the process in which human beings interact with the world and, because of this, continuously define themselves and the world. However, education is essential to guide this human development and foster the ability of people to make moral decisions (Løvlie & Standish, 2002, p. 318).

American pragmatist philosophers Ralph Waldo Emerson, John Dewey and George Herbert Mead adapted these ideas through the twentieth century, building the notion of personal development (Biesta, 2012; Kovalainen, 2012). John Dewey, in particular, argued that education is the 'laboratory' in which philosophical notions are concretized. He saw the need to learn from experience as a combination of passive habituation and intellectual thinking (Juuso, 2012, p. 237) guided by a teacher who draws students' attention to past experiences, tradition and curriculum. Learning, he wrote, is an unending transformation that requires give-and-take communication with others in a community (Juuso, 2012, p. 239). Mead took this a step further, pointing out that meanings cannot be handed down to the learner, but arise through reactions in the mind of the learner, a social process often referred to as 'negotiating meaning', confirmed by empirical classroom observations (Brown, 1992).

The philosophy of *Bildung* in Europe, transformed by Dewey and Mead in the USA, in parallel with the influence of social psychologist Lev Vygotsky, have greatly influenced our understanding of education and have shaped teacher education in recent decades (Bransford et al., 2000; Brown et al., 1993; Pikkarainen, 2014; Sjöström & Eilks, 2018). At the same time, however, the desire by governments to seek economies and efficiencies from their education investments have stimulated contested views of education priorities. Around the 1990s, particularly in the English-speaking world, a view emerged that what students can do as a result of their education is more important than simply acquiring knowledge, and this led to the widespread adoption of educational competencies as markers of attainment, particularly in higher and vocational education. Several writers have pointed out philosophical and educational flaws in this approach (Barnett, 1994, 2009; Hager, 2004; Pikkarainen, 2014), so far without much effect on education practices. In some ways, these writers have echoed Nietzsche's critiques in the 1870s and Adorno a century later who both argued that linking Bildung with economic performance or individual market power and ability is antithetical to its essential notion of moral education (Konrad, 2012, p. 107; Thompson, 2005, p. 521; Wischmann, 2018).

Today, both education philosophy generally and ideas of *Bildung* in particular continue to be contested spaces as contemporary thinking evolves, and issues concerning race, gender, discrimination and sustainability require many fundamentals to be reconsidered (see for example Sjöström et al., 2017; Thompson, 2005; Wischmann, 2018). However, very few engineering educators have even heard of *Bildung* and there are very few accounts like those of Avery and Wihlborg (2013) that explain how *Bildung* influences classroom practices in higher education. Roland Reichenbach (2014) recited Georg Bollenbeck's critique, that scattered remnants of educated middle classes and their semantics remain in philosophy faculties under institutional protection, but outside it is hard to discern any influences in disciplines such as engineering.

Reading just a small selection of literature on the topic reinforces ideas advanced in discussions on *Bildung* in favour of liberal education, intellectual freedom, and moral education that are highly relevant for engineering education. Of course, many writers have advanced similar ideas before, arguing that scientists (and engineers by extension) need these ideas to deal with issues concerning ethics, social justice and sustainability that arise because of the ways that engineering artefacts and systems influence human behaviour at an individual level and even whole societies at a national or even global level (e.g. Bucciarelli, 2008; Bucciarelli & Drew, 2015; Claris & Riley, 2012; Sjöström & Eilks, 2018; Sjöström et al., 2017; Williams, 2003). Guldborg and Skelmose (2023) recently argued that *Bildung* philosophies lead to the contemporary notion of 'action competencies', and that educators can cultivate agency among their engineering students to actively participate in democratic decision-making within a society to improve sustainability outcomes, echoing some of Nussbaums ideas mentioned earlier. Giuliano et al. (2022) argued that engineering judgement requires selection of an appropriate rationality to evaluate the relative value of different engineering actions needed to achieve a given end-state, and that it requires critical thinking, thus justifying the need for humanities to frame an engineering curriculum, presumably philosophy. Anders Buch, however, has cautioned the idealism behind such advocacy arguing that workplace cultures and infrastructures largely attenuate the influences of these progressive pedagogies (Buch, 2015, 2016).

In the rest of this chapter, I take a different approach, arguing that the relevance of *Bildung* and its derivatives emerges from a fundamental reassessment of engineering in practice which is different to what many in the academy imagine it to be. I argue that misunderstandings about engineering practice by educators, imparted to their students, contribute to costly engineering failures that are occurring all around us, yet mostly out of sight and the mind's eye. Earlier writers have advocated structuring education for engineers around ideas from *Bildung* philosophy, or locating humanities and liberal education at the core of engineering education, in order for students to appreciate their ethical responsibilities to society as a whole. Instead, I will argue that these ideas can better prepare students to understand engineering workplace practices, and to help students develop a stronger sense of identity and agency as engineers, and to understand why these costly engineering failures occur, and possibly how such failures could be prevented in future.

In in the next section, I will briefly outline contemporary engineering failures and what we know about their causes.

Following that, I will explain how research on engineering practices is revealing that engineering practice is largely a social discipline in which the solitary cognitive performances such as critical thinking, design, technical problem-solving and calculation comprise relatively minor parts of engineering practice, even though they are still important.

Current engineering education curricula mainly emphasize solitary technical cognitive tasks that comprise only a small but important component of engineering practice. Only a tiny number of engineering educators have sufficient contextual knowledge of the commercial engineering environments in which their students will find employment to explain to their students how the knowledge and skills they learn will be applied in practice. Therefore, there is a large mis-match between the curriculum and practice which most educators cannot bridge. This explains why attempts to frame a new curriculum around the social realities of practice are rarely successful, and tend to be resisted by most engineering faculty.

Therefore, in order to make progress, there needs to be a focus on helping engineering educators understand the practices that their students will be expected to perform. Only when educators themselves understand that engineering is a social system at the core, just as education itself is a social system, will it be possible to introduce ideas like the *Bildung* philosophies we have discussed in this book.

11.2 Contemporary Engineering Failures

A few engineering failures are spectacularly visible, perhaps with a video captured by chance on a mobile phone and widely disseminated.[1] However most are hidden, at least in part to avoid embarrassing the owners. According to industry data (Merrow, 2011; O'Brien, 2009; Young, 2012), most large engineering projects (over USD 1 billion) fail to provide at least half the promised financial return for investors, and about one in six are complete write-offs in which investors lose practically everything. Flyvbjerg (2014) has presented similar evidence. While smaller projects succeed more often, if cars failed at comparable rate, few people would buy them. At the root of most of these failures lie collaboration weaknesses and communication failures. Often the likelihood of failure is known early on, but the project continues in the belief that failure can be avoided. Nearly all of these projects involve well-established technologies that have proved to be successful in earlier ventures: Trevelyan (2014, Ch 11) presents a typical case study. Anecdotal evidence collected by the author from conversations with investors and financiers suggests that these failure rates are appreciated, and financial expectations are appropriately discounted in response. However, the enormous waste of resources, intolerable in other aspects of human life, has to be considered a very large engineering failure. Research interviews by the author help to reveal that many engineers are unaware of these failures, and even if they are, attribute them to "management failures". Yet almost all these projects are managed by experienced engineers.

Another contemporary engineering failure lies almost entirely in the Global South. Engineered goods, artefacts, services, systems and infrastructure, of equivalent quality, design, performance, and durability are considerably more expensive in the Global South than in higher-income countries. While construction is typically 60% more expensive in South Asia than Australia, for example, electricity at the point at which it performs useful work is typically three to five times more expensive. Safe drinking water is ten to thirty times more expensive because, almost invariably, it has to be carried in containers to consumers. Water supply pipe networks are permanently contaminated with faecal bacteria and cannot be relied on to provide safe drinking water, at times providing no water at all. As a direct consequence, the incidence of stunting caused by permanent damage to the human

[1] See, for example, two Indian engineering disasters in a week: https://wp.me/p51WXF-4aK, and the failure of the first Space-X Starship launch: https://www.youtube.com/watch?v=5l_BX12eDZY

intestines caused by repeated diarrhoea infections before the age of two, affects around 45% of infants in South Asia and Sub-Saharan Africa, resulting in malnutrition even if sufficient food is available. Coelho (2004) and Domal (2010) have provided detailed explanations that show how young engineers emerging from engineering schools are completely unprepared for the realities of practice.

The author's research demonstrates a consistent thread linking all engineering performances world-wide. Engineering is an endeavour in which technical knowledge is used to deliver artefacts, systems, infrastructure and information that enables other people to be more productive, to do more with less effort, time, materials, energy, health risks, environmental disturbances and uncertainty. The relative failure of this endeavour in the Global South, so far, limits productivity to about one fifth the level in higher-income countries, meaning that practically everything needed for human existence is more expensive for the same level of quality and performance (Manyika et al., 2015). This failure, in turn, explains why social and economic development is so slow in the Global South.

There are other areas in which engineering failures continue at levels that common sense would consider to be unacceptable, such as maintenance, but the socio-technical complexity defeats naïve improvement attempts (Gouws, 2014; Russell & Vinsel, 2016; Vinck, 2019). All these failures are socio-technical, so can only be addressed by understanding how the social and technical are deeply and inseparably intertwined in engineering practices. For that, as explained in the next section, it is necessary to reconstruct prevalent ideas about engineering practices.

11.3 Reconstructing Engineering from Practice

Over two decades since 2002, I have researched engineering workplace practices in several countries with help from colleagues and research students. The predominant view among engineering academics is that engineering can be described as technical problem-solving under operational and resource constraints (e.g. Purzer et al., 2018; Quinlan, 2002; Sheppard et al., 2006). Koen (2009) described it slightly differently as "use of heuristics to cause the best change in an uncertain situation within the available resources" (p. 360). These views reflect engineering curricula in universities that predominantly consist of courses in which students learn to analyse the performance of engineered artefacts using mathematical techniques, building largely on knowledge from physical sciences and a small but increasingly significant component from the biological sciences. Unsurprisingly, engineering practitioners draw on the same ideas when describing their roles as 'expert problem-solvers'.

Engineering educators also understand that students also have to be communicators, though the overwhelming emphasis in assessment still reflects the technical course content. In addition, educators acknowledge that students need to be able to work with other people, however the role of communication is still largely seen as an end point in technical analysis in which engineers transfer essential information to others (e.g. Tenopir & King, 2004), a view echoed even by experienced practitioners (e.g. Galloway, 2008).

The realities of engineering practice emerging from research, however, are strikingly different. Data from many studies demonstrates that engineers, even at the start of their careers, spend most of their time interacting with other people (e.g. Tenopir & King, 2004; Trevelyan, 2010; Trevelyan & Tilli, 2008; Williams & Figueiredo, 2010; Youngman et al., 1978). A large component reflects technical coordination practices in which engineers gain the willing and conscientious collaboration of others to contribute skilled and knowledgeable performances to an agreed schedule without resorting to organizational authority (Anderson et al., 2010; Blandin, 2012; Trevelyan, 2007). In particular, Trevelyan (2014) argued that this coordination activity, and social interactions more generally reflect that nature of engineering knowledge which is distributed in the minds of participants (Trevelyan, 2010, 2014, Ch 5). The knowledge acquired by engineering students forms a very small part of the knowledge required in an engineering enterprise, so engineers need to rely on social networks of collaboration to gain access to the knowledge they need for their work, usually by informally engaging the skilled performances of other people.

Engineering knowledge, therefore, emerges from a social network as much as, or even more than a written body of knowledge. Engineers also spend a significant amount of time searching for written information on their own, mainly using internet resources, also books, patents and journals. This too, can be seen as a collaborative social activity dispersed in time in which some prepare information for later access by other people searching for it.

Studies have shown, for example, that technical problem-solving in practice largely relies on social interactions with other people. For example, a study of problem-solving in the automotive industry demonstrated that gaining alignment among different stakeholders on the nature of the problem was the main factor that distinguished success from failure, in essence a multi-party negotiation performance (Itabashi-Campbell & Gluesing, 2013). At the start, some of the stakeholders argued that the problem lay with others to avoid having to adjust their work commitments. Jonassen et al. (2006) also identified the collaborative nature of industrial problem-solving. Korte et al. (2008) found that early-career engineers soon learned to find people who knew how to solve a particular problem and that required an understanding of the social dynamics of engineering work.

Classifying engineering performances using well-established occupational frameworks is one way to explain how practice differs significantly from contemporary understandings among engineering academics.

Occupational sociologists classify job tasks in three broad categories (Howell & Wolff, 1991; Spenner, 1990; Zuboff, 1988):

1. *motor skills* requiring manual dexterity and motor coordination;
2. *interactive skills*, such as interpersonal skills, organizational and managerial skills (leadership, autonomy, and responsibility); and

3. *cognitive skills* including verbal and language skills, diagnostic skills (synthetic reasoning abilities, and analytical skills including mathematical and logical reasoning abilities).

From this, we describe three groups of performances evident in extant studies of engineering work.

Group 1: Physical work equivalent to motor skills in which people use expert knowledge to ***interact physically with artefacts and tools***, often referred to as 'hands-on' work. Group 1 performances can be solitary or collaborative when a single pair of hands and eyes is not sufficient. These performances comprise a substantial part of the work of technicians reported in several studies (e.g. Barley & Orr, 1997; Horning, 2004; Orr, 1996; Winberg, 2021).

Group 2: Cognitive work and skills in which people use expert knowledge to *interact mentally with abstract objects and tools***.** For example, programmers use computer-based tools to create software codes based on abstract mental models of computing machines. Plant and system engineers interact with systems such as power grids, chemical process plants, ships, aircraft, telecommunications networks and other complex artefacts by adjusting settings, as often as not in computer systems that serve as an intermediary communication medium between people and these systems. People accumulate explicit, procedural, and tacit knowledge over years of experience from interacting with these systems. We can include information searching and retrieval performances with this group. Like physical performances, cognitive work is either solitary, or collaborative when the task is beyond the capacity of a single person. Design, construction, and sustainment of process plants, power grids, ships and aircraft, for example, require many people simultaneously engaged in cognitive performances collaborating together.

Group 3: **Interactive work**, *social interactions with other people* to plan, organize, collaborate in, and coordinate physical (group 1) and cognitive (group 2) performances. These interactions may be synchronous (e.g. face to face dialogues, in meetings or by phone or video calling or teleconferencing), relying substantially on oral communication supplemented by non-verbal cues and gestures. They can be in formal meetings, informal workplace settings, or socializing outside the workplace. People also interact asynchronously (email correspondence, text messages, supported by boundary objects (Star & Griesemer, 1989; Whyte et al., 2016) such as images, two- and three-dimensional drawings, digital data) using workflow and project management systems and many other forms of computer-mediated communication. These systems have enabled engineers to work on artefacts so elaborate that the details exceed the memory capacity of people to recall accurately (Whyte et al., 2016).

Professional engineers can be distinguished from other engineering occupations, such as technologists, programmers and technicians, by the responsibility for performance and safety through governance of design (Magarian & Seering, 2021). They can also be distinguished from the two other major occupational groups in engineering—technologists and technicians—by the proportion of time spent on these three different groups of performances. Engineering technicians mainly engage in group 1 performances. Engineering technologists mainly engage in group 2 performances.

When we assess quantitative evidence from a wide range of studies, professional engineering work consists primarily of group 3 social interactions. Solitary cognitive work such as design, modelling calculations and technical problem-solving forms a smaller element of practice, typically 10–20%. Most professional engineers perform little if any hand-on physical manipulation of tools and artefacts.

There is also substantial agreement across the small but growing number of ethnographic studies of professional engineers in different countries, industries and settings on the predominance of social interactions in professional engineering work. Most studies have come from higher-income countries, particularly Australia, USA, and European countries. One of the earliest was a study by Zussman who was motivated to study how middle-class engineers participated in political action as a group (Zussman, 1985). The descriptions of their work were seen at the time as a by-product. Bucciarelli studied design engineers in advanced technology companies, and posited that engineers engage with an 'object world' in their thinking, even though their work substantially relied on social interactions as they engaged in complex negotiations with different stakeholders (Bucciarelli, 1994). Others observed similarly complex social interactions (e.g.Darr, 2000; Minneman, 1991; Sandberg, 2000; Suchman, 2000; Suchman et al., 1999). Vinck and his French colleagues presented a series of studies with design as a theme, observing how engineers conduct negotiations to clarify their requirements. Perhaps more than others, they started to examine everyday engineering practices in a variety of industrial settings. Vinck recently returned to this theme in his work, realizing that many other aspects of engineering such as maintenance and repair work had not been studied in sufficient detail (Vinck, 2019), an observation echoed by Colmellere (2022). Maintenance, in particular, can involve complex interactions between people and computer systems employed to provide some coordination of maintenance, upgrades and repair work, and these interactions provide many unexpected results (Gouws, 2014). Japan's economic ascendancy motivated several comparative studies in the 1980s and 1990s as westerners tried to understand their industrial success. These also provided supporting evidence on social interactions, and also provided some of the earliest evidence on how different social and organizational cultures influence engineering practices (e.g. Button & Sharrock, 1994; Kunda, 1992; Lam, 1996, 1997; Petersen, 2015).

Even though social interactions form the predominant work activity for professional engineers, the engineers themselves do not necessarily acknowledge this, seeing this activity as 'non-technical' or 'administration', and 'not real engineering work', even seeing it as 'interrupting' real engineering work (Bailyn & Lynch, 1983; Perlow, 1999). Trevelyan (2010) also observed how engineers relegate social interactions to a secondary or peripheral status, even avoiding some altogether. This tendency significantly contributes to the engineering failures noted above. Evidence from the author's research interviews helps to demonstrate how the avoidance of design checks and inspection work contributes to project delays and cost overruns, even distorting data used to justify major project investment decisions. It is now well understood that avoiding major accidents in complex technological systems requires close attention to socio-technical factors (Busby et al., 2004; Reason & Hobbs, 2003).

Practically all the studies on engineering practices have come from higher-income countries. The few studies that have been conducted from the 'Global South' provide similarly strong supporting evidence for the critical influences of social interactions on workplace practices. In an insightful study of government engineers in Madras (Chennai), Coelho (2004) observed how engineers in a government water utility adapt to, and work around local political power structures, engaging in what many would see as corrupt practices out of sheer necessity to retain their jobs and support families. Domal (2010) compared Indian and Australian private sector metal-manufacturing practices and observed how wide disparities in social power in India ultimately resulted in confusion and uncertainty on the factory floor, causing work to stop in the absence of supervisors. Trevelyan (2022) provided further observations on the effects of South Asian social culture on engineering practices, explaining how these effects significantly raise the cost of engineering work. He also showed how a few engineers develop their own "culture islands" within their enterprises enabling significantly higher levels of performance. Matemba (2020) and (Kusimo & Sheppard, 2019) have provided similar observations from Sub-Saharan Africa.

Trevelyan (2014) described a series of collaborative performances that characterize professional engineering practices. At the lowest level, informal discussions take place that can be described as mutual learning and teaching and learning, where participants acquire knowledge from each other and, in doing so, can also generate new knowledge and insights resulting entirely from their interactions with each other. Some engineers also engage in formal teaching performances. One-one teaching performances are significant in firms using well-established practices that early-career engineers need to learn (Bailey & Barley, 2010). Advocacy and persuasive performances are also significant (Coso Strong et al., 2022) when engineers argue for resources or a particular interpretation of evidence, a frequent performance when engaging with regulatory agencies. Technical coordination performances exhibit a higher level of complexity, consisting of informal learning and teaching, combined with advocacy and persuasion and negotiation as well. These also extend in time as informal follow-up reviews are critical at appropriate time intervals to avoid disappointment. Their formal equivalent, project management, exhibits similar but much more extensive performances supplemented with written documentation and technical specifications. The latter are critical in engineering practice yet almost completely overlooked by educators. Trevelyan (2014) placed multi-party negotiations at the upper end of the socio-technical complexity scale, noting their importance in multi-disciplinary technical problem-solving (Itabashi-Campbell & Gluesing, 2013) and sustainability.

When these complex socio-technical performances are examined closely, we can appreciate that the simplistic information-sharing models of communication typically learned by engineering students (e.g. Tenopir & King, 2004) are insufficient for navigating the realities of practice. However, this is not the only critical weakness in engineering schools today.

11.4 Reconstructing Engineering Education

All the teaching in medical schools beyond the basic sciences, an introduction to pharmacology, and social and population sciences, relies on instructors who practice medicine daily. On the other hand, few if any engineering faculty have contemporary experience of the engineering practices that their students will soon be expected to perform. While there have been no follow-up studies that expand on an Australian survey showing minimal industry experience among engineering faculty (Hadgraft & Cameron, 2010), anecdotal evidence suggests that their results can safely be generalized globally. In the absence of practice experience, engineering faculty develop notions of engineering based on the curriculum they teach, framed in terms of technical problem solving, and consisting largely of scientific and mathematical analysis, with some notional studies in design (e.g. Koen, 2009; Purzer et al., 2018; Quinlan, 2002; Sheppard et al., 2006). Students develop similar notions under the guidance of their educators, and experience surprise, and often dismay when they encounter the realities of practice later. Socialization difficulties are commonly experienced by early-career engineers (e.g. Beddoes, 2021; Brunhaver et al., 2021; Gewirtz et al., 2018; Gewirtz & Paretti, 2021; Jesiek et al., 2021; Johri, 2012; Korte, 2009; Korte et al., 2008; Kovalchuk et al., 2017; Lee, 1986, 1994; Lutz & Paretti, 2021; Petersen, 2015; Trevelyan & Tilli, 2008; Zhu et al., 2021).

Given this background, it is hardly surprising that practicing engineers tend to see socio-technical interactions as ‘not real engineering’ (Trevelyan, 2010). Faculty beliefs (Quinlan, 2002) also help to explain the resistance encountered by faculty who attempt to raise the need for education to address socio-technical aspects of engineering practice in engineering schools. This is a similar phenomenon to the resistance that Quinlan (2002) observed within engineering faculties against a minority who advocate a central place for creative engineering design by the majority who place analysis and technical problem-solving at the core. Xavier et al. (2022) reported comments such as “that social science garbage” from engineering faculty in a study exploring how social aspects of engineering are interpreted during accreditation evaluations. This author has also experienced this resistance, partially explained by a chance conversation with a faculty colleague. She had just learned of another (female) colleague’s commercial engineering experience and remarked “She’s a real engineer!” This comment reveals an identity conflict alluded to by Williams (2003) where engineering faculty identify themselves as engineers within a university community of scholars, even though they suspect that they would not have the knowledge required to practice as an engineer outside the academy. Discussions on the relevance of teaching for industrial practice are likely to surface aspects of this identity conflict. In the experience of this author, most engineering academics experience this conflict at some level.

Most engineering educators believe that by assigning students to groups to perform more complex problem-solving and design tasks, the students will somehow learn collaboration skills through practice (Sheppard et al., 2009) even though observations reveal anti-collaborative practices by students working in groups (e.g. Leonardi et al., 2009).

The lack of contemporary practice experience among engineering faculty, particularly when coupled with resistance to addressing the socio-technical aspects of engineering, provides powerful explanations for the socio-technical system failures noted in the earlier section of this chapter. Observations showing that collaboration weaknesses in higher education propagate into workplaces (Leonardi et al., 2009) further underline the opportunities for educators to address socio-technical aspects of engineering.

The lack of industry experience is certainly evident to early-career engineers interviewed recently by Beddoes (2023). She reported that early-career engineers advocated for engineering faculty to gain practice experience so they could prepare students for the tasks they would be expected to perform as newcomers. Faculty would also be able to explain how classroom learning would be used in practice. These observations lead to the obvious question, how can engineering faculty teach effectively without understanding how their students are expected to perform as newcomers? The argument of this chapter, that the weak understanding of socio-technical aspects of engineering practice leads to repeated and costly engineering failures, might be questioned by some readers, yet the question in the preceding sentence still demands an answer.

In lower-income countries such as India, engineering education weaknesses can be described as 'tragic' with only a small proportion of graduates considered to be employable in engineering firms (Tilak & Choudhury, 2021). The author has witnessed similar issues in other lower-income countries, and detailed workplace studies have provided explanations (Coelho, 2004; Domal, 2010). Given this scenario, how might one begin to restructure engineering education and address contemporary engineering practices, particularly the socio-technical complexities described in the previous section of this chapter? Many engineering faculty might argue that it is sufficient to focus attention on communication skills in the context of technical courses (e.g. Lievens, 2023). However, as Sheppard et al. (2009) observed, faculty acknowledge the importance of social skills, but believe that practice is sufficient for students to gain proficiency. Indeed, there is a common belief among engineers that communication skills can only be learned by practice.

Several have argued for restructuring engineering education by including humanities and the liberal arts in the engineering curriculum. We should note that the Grinter review of engineering (1955) argued for a major social science and humanities curriculum component many decades ago, a recommendation adopted by only a small handful of universities. However, even in the few instances where engineering programs include a significant humanities component, this author is not aware of any that address the research literature on engineering practice that has become available in recent decades.

Most arguments for including humanities focus on the impact that engineering activities have on human societies. The institutions that do include a humanities component justify that with similar arguments. However, as explained in the earlier section of this chapter, the socio-technical complexities of engineering practices and the high incidence of failures caused by collaboration weaknesses provide a far more compelling argument, particularly for students who prefer instruction that is relevant for industry practices. Arguments based on the need for students to acquire critical thinking capabilities (e.g. Giuliano et al., 2022) advocating a humanities core in engineering education might be much stronger with the understanding that engineering is a series of complex social-technical performances that need to be understood to avoid failures.

Perhaps one should start with the observation by Buch (2022) who argued for the "recovery of engineering" to be seen first and foremost as a problem-solving discipline that seeks to ameliorate living conditions. This author has argued in a similar way earlier in this chapter that engineering, at its core, aims to improve human productivity, extending human capabilities.

One way to start might be a graduate diploma or masters level course on engineering practices starting with entry level courses in humanities such as philosophy, social sciences such as industrial anthropology, psychology and communication studies. Most people immersed in an engineering community face steep learning experience because of some deeply engrained assumptions. Engineers (and engineering students) tend to believe that their mathematical fluency confers an ability to think logically that is superior to other people, even other scholarly disciplines. They tend to see argumentation with words as fuzzy and imprecise, indeed this assumption underpins a significant branch of control engineering based on 'fuzzy logic' (Trevelyan, 2014, Ch 3). The way that engineering faculty present lectures reinforces this preference for reliance on mathematical logic. Whereas the mathematical analysis for a textbook problem will be written line by line by the instructor, the assumptions, the logical argument that provide the foundations for the mathematics, and the deductive or inductive steps that link aspects of the problem with particular mathematical equations tend to be spoken and not written out. A further assumption that can create a significant learning challenge is that words have precise definitions. Engineering texts and standards often spell out these definitions as if they are the absolute truth, and many engineers would have difficulty accepting the idea that one can engage in a productive discussion without agreement on the meanings of words.

Taking the need to keep the cost as low as possible, students would join entry-level humanities and social science courses. However, additional readings that link their studies with engineering practice research would help student engagement and keep tutorial discussions relevant for engineers. Ideally, non-engineering students would also read some of the engineering practice research studies as it would change their view of engineering as well. Beyond the foundation, course material addressing specifications, inspection and testing will anchor the programme in practical engineering issues that can be explored with deeper knowledge of human relationships and motivations. For example, experienced engineers can explain how quality control in engineering work depends significantly on the physical presence of external stakeholders. The programme should also include an introduction to engineering systems, the organisational procedures and computer systems that enact large-scale collaboration in engineering. Optional courses could include financial decision-making in engineering, foreign languages and more advanced philosophy studies such as axiology. Role playing simulations can also help students develop a deeper understanding of collaborative performances such as multi-party negotiations.

Ultimately, integrating studies of the socio-technical aspects of engineering in the undergraduate curriculum would be essential to ensure that these aspects are seen as part of the core intellectual foundation of the discipline.

No such restructuring would be possible without engineering faculty acquiring contextual knowledge of engineering practices as a largely social collaborative endeavour. While first-hand experience is one way for faculty to acquire contextual knowledge, it is not the only way. In fact, first-hand experience without appropriate guidance can be just as misleading as the contemporary notions of engineering reported above. This difficulty has been reported in studies of engineering internships. For example Ajjawi et al. (2020) reported that students had learned "communication skills, problem solving and collaboration with others in the workplace, yet expressed frustration at the nature of the assessment tasks being individualistic and focused on a narrow genre of report writing or reflective writing which did not enable them to represent what they had achieved. Drawings reflected this where placement learning activities showed many people, often smiling, communicating and interacting; whereas assessment was represented as the student, alone sitting facing a computer." Here we see how the faculty view of engineering practice as a solitary cognitive activity conflicts with the reality experienced by the internship students who were being assessed.

We could return to the observation of Davis (1998) who argued at the time that we need more knowledge on what engineers actually do. 25 years later, we have a growing body of evidence from what is still relatively small collection of research reports. However, it is quite unreasonable to demand that busy engineering faculty read even a small proportion of these reports when their career prospects depend on building their own research interests. Teaching-only faculty in engineering colleges and schools face demanding classroom teaching schedules and, in many countries, low salaries which they have to supplement by after-hours tutoring.

Story-telling could also be a way for engineering faculty to acquire contextual knowledge they need. Perhaps, then, the task for engineering practice researchers is not only to observe practicing engineers, and report their results in research journals, but also to write short stories that provide compelling and highly readable accounts of early-career engineers and the workplace challenges they face.

Ultimately, however, it is likely that most engineering faculty will only start to reassess their notions of engineering practice with outside pressure from the principal external stakeholders. Both governments and private firms are exposed to the financial losses from failed engineering projects. Until now, however, the fear of being blamed by voters and shareholders for these failures has restricted public disclosures of many failures except for a spectacular few that cannot be hidden. Certainly, the extent of engineering project failures is not widely appreciated currently.

In the Global South where few engineering graduates obtain employment in line with their aspirations, families have the potential to exert pressure for reform. Another significant external influence comes from accreditation agencies such as the Accreditation Board for Engineering and Technology (ABET) in the USA. However, strong arguments for education reform based on a large number of research studies have made little observable impact on accreditation criteria (e.g. Passow & Passow, 2017), partly because few if any of the faculty who serve on accreditation review panels have read the research. Lastly, by a process of slow knowledge transfer, results from engineering practice research studies have the potential to gradually influence engineering faculty views. However, there are few signs that this change is gathering momentum. For the time being, therefore, it is safe to assume that engineering education will change slowly if at all. For example, an Australian review setting out the scope of changes needed to meet the expectations of the profession in 2035 remarked that resource constraints and the lack of incentives for faculty to learn new pedagogies were significant barriers for change (Crosthwaite, 2021).

Even if education can be restructured, societal and organisational cultural practices combined with the infrastructures that support engineering practice exert strong influences that seem to defeat even the most progressive education interventions (Buch, 2016; Petersen & Buch, 2016). Currently there is almost no research-based understanding on how engineering education influences engineering practices. Therefore, new research directions will be essential to resolve these issues, perhaps more firmly based on contemporary practice research methods.

11.5 Conclusion. The Relevance of *Bildung* in Engineering Education

In the meantime, it is possible to learn from observations on engineering practice related to the consequences of an education system focusing primarily on the technical knowledge required by engineers. We can observe how education practices shape the expectations and habits of engineers and constrain their ability to perform in practice.

Informal technical coordination practices in engineering appear to rely strongly on trust and referent power (Norman, 2019; Raven, 1992), even though engineers still may not realise this after decades of experience (Trevelyan, 2007). Therefore, personal and character development in the context of social culture are likely to strongly influence engineers' workplace practices, particularly in lower-income countries (Trevelyan, 2022). Even though there seem to be few references to Bildung and its influence on classroom practices in higher education (Avery & Wihlborg, 2013; Sjöström et al., 2017, p. 175), the strong connection between workplaces coordination practices, relationships and character development seems to offer a fertile niche for Bildung philosophical ideas to gain a place in engineering education.

A wide range in personal development and character can be expected among engineers as a group, even though some firms may have strong personality filtering in their recruitment processes. The tendency to see social interactions as 'not real engineering work' or as interruptions may delay or slow the development of social skills among engineers. Naturally, psychological factors such as the autism spectrum disorder (ASD) can also influence personality and social interactions. While there have been discussions on the prevalence of autism among engineers and other STEM occupations, there is no strong evidence of higher participation rates by people on the autism spectrum (Baron-Cohen et al., 2001; Wei et al., 2013) even though it is widely thought that such individuals have a preference for STEM occupations.

The realisation that engineering relies on small- and large-scale collaboration practices based on specialized technical knowledge opens up new spaces for discussions of Bildung philosophy in engineering education. The slow process of education transformation in engineering faculties means that there is plenty of time for careful consideration and discussions on how best to build on two centuries of philosophical development that has brought us to this discussion today.

Acknowledgements The research that led to this chapter was only possible because of the willing participation by hundreds of engineers in the research studied that provide the ideas presented here. Many of the research interviews and field studies were performed by the author's students. In addition, many fruitful discussions with Bill Williams, Russell Korte, Sabbia Tilli and Sally Male helped develop the ideas in this chapter. The author also acknowledges the support from The University of Western Australia and financial support from companies and generous donors.

Conflict of Interest The author has not disclosed any conflicts of interest.

References

Ajjawi, R., Tai, J., Huu Nghia, T. L., Boud, D., Johnson, L., & Patrick, C.-J. (2020). Aligning assessment with the needs of work-integrated learning: The challenges of authentic assessment in a complex context. *Assessment and Evaluation in Higher Education, 45*(2), 304–316. https://doi.org/10.1080/02602938.2019.1639613

Anderson, K. J. B., Courter, S. S., McGlamery, T., Nathans-Kelly, T. M., & Nicometo, C. G. (2010). Understanding engineering work and identity: A cross-case analysis of engineers within six firms. *Engineering Studies, 2*(3), 153–174. https://doi.org/10.1080/19378629.2010.519772

Avery, H., & Wihlborg, M. (2013). Teachers' interpretation of Bildung in practice: Examples from higher education in Sweden and Denmark. *Journal of Learning Development in Higher Education, 2013*(5) https://urn.kb.se/resolve?urn=urn:nbn:se:hj:diva-21367

Bailey, D. E., & Barley, S. (2010). Teaching-learning ecologies: Mapping the environment to structure through action. *Organization Science Articles in Advance*, 1–25. doi:https://doi.org/10.1287/orsc.1090.0511.

Bailyn, L., & Lynch, J. T. (1983). Engineering as a life-long career: Its meaning, its satisfactions, its difficulties. *Journal of Occupational Behaviour (pre 1986), 4*(4), 263–283.

Barley, S. R., & Orr, J. (Eds.). (1997). *Between craft and science: Technical work in US settings*. Cornell University Press.

Barnett, R. (1994). *The limits of competence: knowledge, higher education and society*. Open University Press.

Barnett, R. (2009). Knowing and becoming in the higher education curriculum. *Studies in Higher Education, 34*(4), 429–440. https://doi.org/10.1080/03075070902771978

Baron-Cohen, S., Wheelwright, S., Skinner, R., Martin, J., & Clubley, E. (2001). The autism-spectrum quotient (AQ): Evidence from asperger syndrome/high-functioning autism, males and females, scientists and mathematicians. *Journal of Autism and Developmental Disorders, 31*(1), 5–17. https://doi.org/10.1023/A:1005653411471

Beddoes, K. (2021). Examining privilege in engineering socialization through the stories of newcomer engineers. *Engineering Studies, 13*(2), 158–179. https://doi.org/10.1080/19378629.2021.1958824

Beddoes, K. (2023, January 3–6). *Early career engineers' advice for civil engineering instructors hawaii international conference on education*, Honolulu. https://www.researchgate.net/profile/Kacey-Beddoes/publication/367272871_Early_Career_Engineers'_Advice_for_Civil_Engineering_Instructors/links/63c9ce64d9fb5967c2ecda6f/Early-Career-Engineers-Advice-for-Civil-Engineering-Instructors.pdf

Biesta, G. (2012). George Herbet Mead: Formation through communication. In P. Siljander, A. Kivelä, & A. Sutinen (Eds.), *Theories of Bildung and growth: Connections and controversies between continental educational thinking and American pragmatism* (pp. 247–260). SensePublishers. https://doi.org/10.1007/978-94-6209-031-6_12

Blandin, B. (2012). The competence of an engineer and how it is Built through an apprenticeship program: A tentative model. *International Journal of Engineering Education, 28*(1), 57–71.

Bransford, J. D., Brown, A. L., Cocking, R. R., Donovan, M. S., & Pellegrino, J. W. (Eds.). (2000). *How people learn: Brain, mind, experience and school*. National Academy Press.

Brown, A. L. (1992). Design experiments: Theoretical and methodological challenges in creating complex interventions in classroom settings. *The Journal of the Learning Sciences, 2*(2), 141–178.

Brown, A. L., Ash, D., Rutherford, M., Nakagawa, K., Gordon, A., & Campione, J. C. (1993). Distributed expertise in the classroom. In G. Salomon (Ed.), *Distributed cognitions: Psychological and educational considerations* (pp. 188–228). Cambridge University Press.

Brunhaver, S. R., Jesiek, B. K., Korte, R. F., & Strong, A. C. (2021). The early career years of engineering: Crossing the threshold between education and practice. *Engineering Studies, 13*(2), 79–85. https://doi.org/10.1080/19378629.2021.1961570

Bucciarelli, L. L. (1994). *Designing engineers*. MIT Press.

Bucciarelli, L. (2008). Ethics and engineering education. *European Journal of Engineering Education, 33*(2), 141–150. https://doi.org/10.1080/03043790801979856

Bucciarelli, L. L., & Drew, D. E. (2015). Liberal studies in engineering – A design plan. *Engineering Studies, 7*(2–3), 103–122. https://doi.org/10.1080/19378629.2015.1077253

Buch, A. (2015). How progressive pedagogics lie. *Engineering Studies, 7*(2–3), 171–173.

Buch, A. (2016). Ideas of holistic engineering meet engineering work practices. *Engineering Studies, 8*(2), 140–161.

Buch, A. (2022). The need for a recovery of engineering. In S. H. Christensen, A. Buch, E. Conlon, C. Didier, C. Mitcham, & M. Murphy (Eds.), *Engineering, social sciences, and the humanities* (Vol. 42, pp. 361–374). Springer.

Busby, J. S., Hibberd, R. E., Mileham, A. R., & Mullineux, G. (2004). Failure modes analysis of organizational artefacts that protect systems. *Proceedings of the Institution of Mechanical Engineers Part B: Engineering Manufacture, 218*(9), 1211–1215.

Button, G., & Sharrock, W. (1994). Occasioned practices in the work of software engineers. In M. Jirotka & J. A. Goguen (Eds.), *Requirements engineering: Social and technical issues* (pp. 217–240). Academic Press.

Claris, L., & Riley, D. (2012). Situation critical: Critical theory and critical thinking in engineering education. *Engineering Studies, 4*(2), 101–120. https://doi.org/10.1080/19378629.2011.649920

Coelho, K. (2004). *Of engineers, rationalities and rule: An ethnography of neoliberal reform in and urban water utility in South India* [PhD, University of Arizona]. Tucson. http://hdl.handle.net/10150/280681.

Colmellere, C. (2022). Research on engineers' work at a turning point? In S. H. Christensen, A. Buch, E. Conlon, C. Didier, C. Mitcham, & M. Murphy (Eds.), *Engineering, social sciences, and the humanities: Have their conversations come of age?* (pp. 111–131). Springer International Publishing. https://doi.org/10.1007/978-3-031-11601-8_6

Coso Strong, A., Stockman, T., Heale, T., Meyer, S., & Meyerson, E. (2022). Persuasive communication practices of engineers in cross-boundary decision-making. *Engineering Studies, 14*(3), 216–238. https://doi.org/10.1080/19378629.2022.2144738

Crosthwaite, C. (2021). *Engineering futures 2035: Engineering education programs, priorities and pedagogies.* https://www.aced.edu.au/downloads/Engineering%20Futures%202035%20R2%20report%20to%20ACED.pdf

Darr, A. (2000). Technical labour in an engineering boutique: Interpretive frameworks of sales and R&D engineers. *Work, Employment and Society, 14*(2), 205–222.

Davis, M. (1998). *Thinking like an engineer.* Oxford University Press. http://www.amazon.com/Thinking-Like-Engineer-Profession-Professional/dp/0195120515#reader_0195120515

Domal, V. (2010). *Comparing engineering practice in South Asia and Australia* [PhD thesis, The University of Western Australia]. Perth. http://research-repository.uwa.edu.au/files/3224816/Domal_Vinay_2010.pdf

Flyvbjerg, B. (2014). What you should know about megaprojects and why: An overview. *Project Management Journal, 45*(2), 6–19. https://doi.org/10.1002/pmj.21409

Galloway, P. (2008). *The 21st century engineer: A proposal for engineering education reform.* ASCE.

Gewirtz, C., & Paretti, M. C. (2021). Becoming after college: Agency and structure in transitions to engineering work. *Engineering Studies, 13*(2), 111–131. https://doi.org/10.1080/19378629.2021.1959597

Gewirtz, C., Kotys-Schwartz, D. A., Knight, D., Paretti, M., Arunkumar, S., Ford, J. D., Howe, S., Rosenbauer, L. M., Alvarez, N. E., & Deters, J. (2018, June 24–27). New engineers' first three months: A study of the transition from cap-stone design courses to workplaces. *American society for engineering education annual conference*, Salt Lake City, Utah.

Giuliano, H. G., Giri, L. A., Nicchi, F. G., Weyerstall, W. M., Ferreira Aicardi, L. F., Parselis, M., & Vasen, F. (2022). Critical thinking and judgment on engineer's work: Its integration in engineering education. *Engineering Studies, 14*(1), 6–16. https://doi.org/10.1080/19378629.2022.2042003

Gouws, L. (2014). *The extent and nature of problems encountered in maintenance: The story behind the text* [PhD thesis, The University of Western Australia]. Perth. http://research--repository.uwa.edu.au/files/3407587/Gouws_Leonie_2014.pdf

Grinter, L. E. (1955). Report of the committee on evaluation of engineering education (Grinter Report). *Journal of Engineering Education, 44*(3), 25–60. http://www.asee.org/member-resources/reports

Guldborg, C., & Skelmose, C. (2023, June 21–23). Action competence in PBL: Revitalizing educational ideals to foster meaning and engagement in engineering education for sustainable development. *9th International research symposium on PBL: Transforming engineering education 2023*, Aalborg.

Hadgraft, R. G., & Cameron, I. (2010). *The engineering and ICT disciplines: Threshold learning outcomes.* Australian Learning and Teaching Council.

Hager, P. (2004). The competence affair, or why vocational education and training urgently needs a new understanding of learning. *Journal of Vocational Education and Training, 56*(3), 409–433. https://doi.org/10.1080/13636820400200262

Horkheimer, M. (1953). Begriff der Bildung. In *Gegenbwärtige Probleme der Univäristät* (pp. 14–15). Vittorio Klostermann.

Horning, S. S. (2004). Engineering the performance: Recording engineers, tacit knowledge and the art of controlling sound. *Social Studies of Science, 34*(5), 703–731. https://doi.org/10.1177/0306312704047536

Howell, D. R., & Wolff, E. N. (1991). Trends in the growth and distribution of skills in the U.S. workplace, 1960–1985. *ILR Review, 44*(3), 486–502. https://doi.org/10.1177/001979399104400306

Itabashi-Campbell, R., & Gluesing, J. (2013). Engineering problem-solving in social contexts: 'collective wisdom' and 'ba'. In B. Williams, J. D. Figueiredo, & J. P. Trevelyan (Eds.), *Engineering practice in a global context: Understanding the technical and the social* (pp. 129–158). CRC/Balkema. https://doi.org/10.1201/b15792-7

Jesiek, B. K., Buswell, N. T., & Nittala, S. (2021). Performing at the boundaries: Narratives of early career engineering practice. *Engineering Studies, 13*(2), 86–110. https://doi.org/10.1080/19378629.2021.1959596

Johri, A. (2012). Learning to demo: The sociomateriality of newcomer participation in engineering research practices. *Engineering Studies, 4*(3), 249–269. https://doi.org/10.1080/19378629.2012.709520

Jonassen, D., Strobel, J., & Lee, C. B. (2006). Everyday problem solving in engineering: Lessons for engineering educators. *Journal of Engineering Education, 95*(2), 139–151.

Juuso, H. (2012). The origins and educational significance of John Dewey's philosophy. In P. Siljander, A. Kivelä, & A. Sutinen (Eds.), *Theories of Bildung and growth: Connections and controversies between continental educational thinking and American pragmatism* (pp. 227–245). SensePublishers. https://doi.org/10.1007/978-94-6209-031-6_12

Koen, B. V. (2009). The engineering method and its implications for scientific, philosophical and universal methods. *The Monist, 92*(3), 357–386.

Konrad, F.-M. (2012). Wilhelm Von Humboldt's contribution to a theory of Bildung. In P. Siljander, A. Kivelä, & A. Sutinen (Eds.), *Theories of Bildung and growth: Connections and controversies between continental educational thinking and American pragmatism* (pp. 124–107). SensePublishers. https://doi.org/10.1007/978-94-6209-031-6_7

Korte, R. (2009). How newcomers learn the social norms of an organization: A case study of the socialization of newly hired engineers. *Human Resource Development Quarterly, 20*(3), 285–306. https://doi.org/10.1002/hrdq.20016

Korte, R., Sheppard, S. D., & Jordan, W. (2008, June 22–26). *A qualitative study of the early work experiences of recent graduates in engineering*. American Society for Engineering Education, .

Kovalainen, H. A. (2012). Emersonian self-culture and individual growth. In P. Siljander, A. Kivelä, & A. Sutinen (Eds.), *Theories of Bildung and growth: Connections and controversies between continental educational thinking and American pragmatism* (pp. 183–197). SensePublishers. https://doi.org/10.1007/978-94-6209-031-6_12

Kovalchuk, S., Ghali, M., Klassen, M., Reeve, D., & Sacks, R. (2017). Transitioning from university to employment in engineering: The role of curricular and co-curricular activities. *American society for engineering education annual conference*, Columbus, OH.

Kunda, G. (1992). *Engineering culture*. Temple University Press.

Kusimo, A. C., & Sheppard, S. D. (2019, Ausgust 18–21). *Manufacturing bowler ability: How resource scarcity hinders team coordination during manufacturing in Nigerian factories* ASME international design engineering technical, conferences and computers and information in engineering conference IDETC/CIE2019, Anaheim, CA. https://asmedigitalcollection.asme.org/IDETC-CIE/proceedings-pdf/IDETC-CIE2019/59278/V007T06A028/6454036/v007t06a028-detc2019-97820.pdf

Lam, A. (1996). Engineers, management and work organization: A comparative analysis of engineers' work roles in British and Japanese electronics firms. *Journal of Management Studies, 33*(2), 183–212.

Lam, A. (1997). Embedded firms, embedded knowledge: Problems of collaboration and knowledge transfer in global cooperative ventures. *Organization Studies, 18*(6), 973–996.

Lee, D. M. S. (1986). Intellectual, motivational, and interpersonal qualities and NOT academic performance predicted performance on first job. *IEEE Transactions on Engineering Management, EM33*(3), 127–133.

Lee, D. M. S. (1994). Social ties, task-related communication and first job performance of young engineers. *Journal of Engineering and Technology Management, 11*, 203–228. https://doi.org/0923-4748(94)00012-3

Leonardi, P. M., Jackson, M. H., & Diwan, A. (2009). The enactment-externalization dialectic: Rationalization and the persistence of counterproductive technology design practices in student engineering. *Academy of Management Journal, 52*(2), 400–420.

Lievens, J. (2023, June 19–22). Myth-busters at work: Development of engineering identity and employability through student research. *9th International conference on higher education advances (HEAd'23)*, Universitat Politecnica de Valencia.

Løvlie, L., & Standish, P. (2002). Introduction: Bildung and the idea of a liberal education. *Journal of Philosophy of Education, 36*(3), 317–340. https://doi.org/10.1111/1467-9752.00279

Lutz, B., & Paretti, M. C. (2021). Exploring the social and cultural dimensions of learning for recent engineering graduates during the school-to-work transition. *Engineering Studies, 13*(2), 132–157. https://doi.org/10.1080/19378629.2021.1957901

Magarian, J. N., & Seering, W. P. (2021). *Characterizing engineering work in a changing world: Synthesis of a typology for engineering students' occupational outcomes*. In (Vol. 110, pp. 458–500).

Manyika, J., Woetzel, J., Dobbs, R., Remes, J., Labaye, E., & Jordan, A. (2015). *Global growth: Can productivity save the day in an aging world?* McKinsey Global Institute. https://www.mckinsey.com/~/media/mckinsey/featured%20insights/employment%20and%20growth/can%20long%20term%20global%20growth%20be%20saved/mgi%20global%20growth_executive%20summary_january%202015.pdf.

Matemba, E. E. (2020). *Globalisation of engineers: African professional engineering education perspectives* (Publication Number http://hdl.handle.net/20.500.11937/82165) Curtin University]. Perth. https://espace.curtin.edu.au/bitstream/handle/20.500.11937/82165/Matemba%20E%202020.pdf?sequence=1&isAllowed=y

Merrow, E. W. (2011). *Industrial megaprojects: Concepts, strategies, and practices for success*. Wiley.

Minneman, S. L. (1991). *The social construction of a technical reality: Empirical studies of group engineering design practice*. Stanford. https://www.proquest.com/docview/303946198/1960A141C05E4CFCPQ/1?accountid=14681

Norman, F. G. (2019). *Working together, apart: Interpersonal communications within virtual team engineering projects in the WA resource industry* [PhD, Curtin University]. Perth. https://espace.curtin.edu.au/bitstream/handle/20.500.11937/76366/Norman%20F%202019.pdf?sequence=1

Nussbaum, M. (2003). Capabilities as fundamental entitlements: Sen and social justice. *Feminist Economics, 9*(2–3), 33–59. https://doi.org/10.1080/1354570022000077926

Nussbaum, M. (2009). Education for profit, education for freedom. *Liberal Education, 95*(3), 6–13.

O'Brien, J. (2009). *Performance of capital projects in Australian processing industries*. I. P. A. Inc. https://projectcontrolsonline.com/images/technical-paper/Performance-of-Projects-in-Australian-Industries.pdf

Orr, J. (1996). *Talking about machines: An ethnography of a modern job*. Cornell University Press.

Passow, H. J., & Passow, C. H. (2017). What competencies should undergraduate engineering programs emphasize? A systematic review. *Journal of Engineering Education, 106*(3), 475–526. https://doi.org/10.1002/jee.20171

Perlow, L. A. (1999). The time famine: Towards a sociology of work time. *Administrative Science Quarterly, 44*(1), 57–81. https://doi.org/10.2307/2667031

Petersen, R. P. (2015). *Practicing and changing engineering design: A practice perspective on engineering design education*. Aalborg University.

Petersen, R. P., & Buch, A. (2016). Making room in engineering design practices. *Engineering Studies, 8*(2), 23. https://doi.org/10.1080/19378629.2016.1200591

Pikkarainen, E. (2012). Signs of reality: The idea of general bildung by J. A. Comenius. In P. Siljander, A. Kivelä, & A. Sutinen (Eds.), *Theories of Bildung and growth: connections and controversies between continental educational thinking and American pragmatism* (pp. 19–29). SensePublishers. https://doi.org/10.1007/978-94-6209-031-6_2

Pikkarainen, E. (2014). Competence as a key concept of educational theory: A semiotic point of view. *Journal of Philosophy of Education, 48*(4), 621–636. https://doi.org/10.1111/1467-9752.12080

Purzer, S., Moore, T. J., & Dringenberg, E. (2018). Engineering cognition: A process of knowledge acquisition and application. In Y. J. Dori, Z. R. Mevarech, & D. R. Baker (Eds.), *Cognition, metacognition, and culture in STEM education: Learning, teaching and assessment* (pp. 167–190). Springer International Publishing. https://doi.org/10.1007/978-3-319-66659-4_8

Quinlan, K. M. (2002). Scholarly dimensions of academics' beliefs about engineering education. *Teachers and Teaching: Theory and Practice, 8*(1), 41–64. https://doi.org/10.1080/13540600120110565

Raven, B. H. (1992). A power/interaction model of social influence: French and Raven thirty years later. *Journal of Social Behavior and Personality, 7*(2), 217–244.

Reason, J., & Hobbs, A. (2003). *Managing maintenance error*. Ashgate.

Reichenbach, R. (2014). Humanistic Bildung: Regulative idea or empty concept? *Asia Pacific Education Review, 15*(1), 65–70. https://doi.org/10.1007/s12564-013-9298-1

Russell, A., & Vinsel, L. (2016). *Hail the maintainers*. Aeon Media Group Ltd. Retrieved September 6 from https://aeon.co/essays/innovation-is-overvalued-maintenance-often-matters-more

Sandberg, J. (2000). Understanding human competence at work: An interpretive approach. *Academy of Management Journal, 43*(1), 9–25.

Sheppard, S. D., Colby, A., Macatangay, K., & Sullivan, W. (2006). What is engineering practice? *International Journal of Engineering Education, 22*(3), 429–438.

Sheppard, S. D., Macatangay, K., Colby, A., & Sullivan, W. (2009). *Educating engineers*. Jossey-Bass (Wiley).

Siljander, P., & Sutinen, A. (2012). Introduction. In P. Siljander, A. Kivelä, & A. Sutinen (Eds.), *Theories of Bildung and growth: Connections and controversies between continental educational thinking and American pragmatism* (pp. 1–18). SensePublishers. https://doi.org/10.1007/978-94-6209-031-6_1

Sjöström, J., & Eilks, I. (2018). Reconsidering different visions of scientific literacy and science education based on the concept of Bildung. In Y. J. Dori, Z. R. Mevarech, & D. R. Baker (Eds.), *Cognition, metacognition, and culture in STEM education: Learning, teaching and assessment* (pp. 65–88). Springer International Publishing. https://doi.org/10.1007/978-3-319-66659-4_4

Sjöström, J., Frerichs, N., Zuin, V. G., & Eilks, I. (2017). Use of the concept of Bildung in the international science education literature, its potential, and implications for teaching and learning. *Studies in Science Education, 53*(2), 165–192. https://doi.org/10.1080/03057267.2017.1384649

Spenner, K. I. (1990). Skill: Meanings, methods, and measures. *Work and Occupations, 17*(4), 399–421. https://doi.org/10.1177/0730888490017004002

Star, S. L., & Griesemer, J. R. (1989). Institutional ecology, 'translations' and boundary objects: Amateurs and professionals in Berkeley's museum of vertebrate zoology, 1907-39. *Social Studies of Science, 19*(3), 387–420. https://doi.org/10.1177/030631289019003001

Suchman, L. (2000). Organising alignment: A case of bridge-building. *Organization, 7*(2), 311–327. https://doi.org/10.1177/135050840072007

Suchman, L., Blomberg, J., Orr, J., & Trigg, R. (1999). Reconstructing technologies as social practice. *The American Behavioural Scientist, 43*(3), 392–408. https://doi.org/10.1177/00027649921955335

Tenopir, C., & King, D. W. (2004). *Communication patterns of engineers*. IEEE Press - Wiley & Sons. https://doi.org/10.1007/978-94-010-0397-1_17

Thompson, C. (2005). The non-transparency of the self and the ethical value of Bildung. *Journal of Philosophy of Education, 39*(3), 519–533. https://doi.org/10.1111/j.1467-9752.2005.00451.x

Tilak, J. B. G., & Choudhury, P. K. (2021). Employment and employability of engineering graduates in India. *Journal of Contemporary Educational Research, 5*(3), 14. https://doi.org/10.26689/jcer.v5i3.1825
Trevelyan, J. P. (2007). Technical coordination in engineering practice. *Journal of Engineering Education, 96*(3), 191–204. https://doi.org/10.1002/j.2168-9830.2007.tb00929.x
Trevelyan, J. P. (2010). Reconstructing engineering from practice. *Engineering Studies, 2*(3), 175–195. https://doi.org/10.1080/19378629.2010.520135
Trevelyan, J. P. (2014). *The making of an expert engineer*. CRC Press/Balkema - Taylor & Francis.
Trevelyan, J. P. (2022). A social science research agenda for engineering practice. In S. H. Christensen, A. Buch, E. Conlon, C. Didier, C. Mitcham, & M. Murphy (Eds.), *Engineering, Social sciences, and the humanities: Have their conversations come of age?* Springer Science+Business Media B.V. https://doi.org/10.1007/978-3-031-11601-8_13
Trevelyan, J. P., & Tilli, S. (2008, June 20–22). Longitudinal study of Australian engineering graduates: Perceptions of working time. *American society for engineering education annual conference*, Pittsburgh.
Vinck, D. (2019). Maintenance and repair work. *Engineering Studies, 11*. https://doi.org/10.1080/19378629.2019.1655566
Wei, X., Yu, J. W., Shattuck, P., McCracken, M., & Blackorby, J. (2013). Science, technology, engineering, and mathematics (STEM) participation among college students with an autism spectrum disorder. *Journal of Autism and Developmental Disorders, 43*(7), 1539–1546. https://doi.org/10.1007/s10803-012-1700-z
Whyte, J., Lindkvist, C., & Jaradat, S. (2016). Passing the baton? Handing over digital data from the project to operations. *Engineering Project Organization Journal, 6*(1), 2–14. https://doi.org/10.1080/21573727.2015.1115396
Williams, R. (2003, January 24). *Education for the profession formerly known as engineering*. The Chronicle of Higher Education, B12.
Williams, B., & Figueiredo, J. (2010, April 14–16). Engineers and their practice: A case study. *IEEE conference on education engineering (EDUCON)*, Madrid.
Winberg, C. (2021). The making of engineering technicians: Ontological formation in laboratory practice. *Engineering Studies, 13*, 226–248. https://doi.org/10.1080/19378629.2021.1987921
Wischmann, A. (2018). The absence of 'race' in German discourses on Bildung. Rethinking Bildung with critical race theory. *Race Ethnicity and Education, 21*(4), 471–485. https://doi.org/10.1080/13613324.2016.1248834
Xavier, P., Wint, N., & Orbaek White, G. (2022). A snapshot of how 'social' considerations are currently being interpreted and addressed within engineering education and accreditation. In S. H. Christensen, A. Buch, E. Conlon, C. Didier, C. Mitcham, & M. Murphy (Eds.), *Engineering, social sciences, and the humanities: Have their conversations come of age?* (pp. 65–92). Springer International Publishing. https://doi.org/10.1007/978-3-031-11601-8_4
Young, R. (2012). *The performance of Australian industrial projects* [Consultant report]. 8. Retrieved 3 May 2016, from https://www.afr.com/companies/manufacturing/planning-failure-a-waste-of-precious-20090910-jn7jr
Youngman, M., Oxtoby, R., Monk, J. D., & Heywood, J. (1978). *Analysing jobs*. Gower Press.
Zhu, J., Hu, Y., Li, Y., Zhang, Z., & Li, W. (2021). Perceptions towards prior learning experiences: Lessons learned from early and mid-career professional engineers in a Chinese context. *European Journal of Engineering Education*, 1–17. doi:https://doi.org/10.1080/03043797.2021.1954602.
Zuboff, S. (1988). *In the age of the smart machine: The future of work and power*. Basic Books.
Zussman, R. (1985). *Mechanics of the middle class: Work and politics among American engineers*. University of California Press.

James P. Trevelyan, Emeritus Professor, engineer, educator, and researcher with 45 years of experience became a start-up entrepreneur in 2015. He leads Close Comfort, inventing and commercializing energy saving, climate friendly Coolzy personal cooling technology. He taught design, mechatronics, project management and sustainability at The University of Western Australia. He is best known internationally for pioneering research that resulted in sheep shearing robots and for the first industrial robot that could be remotely operated via the internet in 1994. He has received international awards for robotics research and engineering education. He also researched landmine clearance methods. In 2002, he was elected a Fellow of Engineers Australia and started researching engineering practice, what engineers really do in their work. He built on this research to design the current Australian stage 2 professional engineering competencies for engineers, technologists, and associates. His 2014 book The Making of an Expert Engineer provides, for the first time, a comprehensive research-based understanding of engineering practice in several countries. It explains why engineered products and services are usually far more expensive in developing countries than the industrialized world, which in turn explains why so many people continue to live in stubbornly persistent poverty.

Chapter 12
Toward a Self-Cultivation Approach to Engineering Education: A Confucian Perspective

Qin Zhu and Yixiang Shawn Sun

Abstract This chapter serves as an initial endeavor to investigate the potential utility of the idea of self-cultivation in constructing an alternative approach to prevalent approaches in engineering education, which predominantly concentrate on the development of employable skills for individual engineers. To achieve this goal, this chapter first conceptualizes a challenge central to dominant approaches to engineering education which is the lack of the self-dimension. It then provides a brieft introduction to the Confucian theory of self-cultivation by comparing it with the Western concept of *Bildung*. Finally, this chapter discusses the implications of this self-cultivation theory for engineering education practice and research. It explores how teaching, learning, and assessment in engineering education can be defined and implemented differently. This chapter is expected to be of particular interest to engineering educators, administrators, and policymakers who are unsatisfied with dominant approaches to engineering education that focus on the attainment of measurable technical knowledge and skills.

Keywords Engineering education · Self-cultivation · Confucian ethics · *Bildung* · Educational theory

12.1 Introduction

Philosophers and engineering educators have recently criticized that dominant approaches to the professional formation of engineers lack self-critical and self-reflective dimensions (Karwat et al., 2015; Karwat, 2019; Mitcham, 2014). While engineering students are taught to comply with professional codes of ethics and be aware of the social and ethical implications of technology (Herkert, 2000), rare

Q. Zhu (✉) · Y. S. Sun
Department of Engineering Education, Virginia Tech, Blacksburg, VA, USA
e-mail: qinzhu@vt.edu; shawnsun@vt.edu

A. Buch, S. H. Christensen (eds.), *Bildung for Engineering Education and Practice*, Philosophy of Engineering and Technology 49,
https://doi.org/10.1007/978-3-031-86581-7_12

opportunities can be found in the curriculum that allow them to reflect on some fundamental and self-reflective questions for the engineering profession such as why they are engineers, for whose benefit they work, and what kinds of engineers they want to become (Zhu, 2021). By drawing on the theories in Confucian ethics (a philosophical tradition known for its emphasis on the value of self-cultivation for human development and social governance), this chapter aims to develop an outline for the self-cultivation approach to engineering education practice and research.

Such a process encompasses two components. First, this chapter conceptualizes a challenge central to dominant approaches to engineering education which is the lack of the self-dimension. Second, it will briefly introduce the Confucian theory of self-cultivation by comparing it with the Western concept of *Bildung*. Conducting such a cross-cultural comparison is meaningful in at least two different ways. First, such a comparison allows readers to develop a much deeper and critical understanding of Confucian self-cultivation including its underlying values, commitments, and limitations. Second, doing so will also be helpful for further augmenting Confucian self-cultivation by integrating insights from the Western interpretations of *Bildung*. Therefore, a self-cultivation theory developed in this way will be more responsive to diverse cultures. Even though such a theory was derived from the Confucian tradition, it has been critically examined and expanded in its comparison with the Western concept *Bildung*.

Finally, this chapter will discuss the implications of this self-cultivation theory for engineering education practice and research. It explores how teaching, learning, and assessment in engineering can be defined and implemented differently. Doing so generates insights that can be potentially useful for designing instructional activities that invite students to reflect on (1) the values that drive their personal and professional decision-making; and (2) how their participation in engineering transforms their professional identities and the values and rituals in communities they serve. By taking the lens of self-cultivation, this chapter also critically examines the typical learning outcomes in engineering. It analyzes how these learning outcomes and the ways in which they are assessed in educational practices may potentially undermine the development of self-cultivation in students. It will also explore how these learning outcomes can be rewritten to include space for students' self-cultivation. This chapter is expected to be of particular interest to engineering educators, administrators, and policymakers who are unsatisfied with dominant approaches to engineering education that focus on the attainment of measurable technical knowledge and skills.

12.2 The Self: A Missing Dimension in Engineering Education

Dominant ideologies in engineering education have persistently been criticized by humanists and social scientists for lacking the self or the critical dimensions. There is a widespread ideology "normative holism" in engineering education that equates "the technical contents of engineering practices with material advancements across the planet for human benefit" (Downey, 2012, p. 233). Partly what this ideology leads to is that engineers do not feel obligated to critically examine the actual consequences of their work including their *unique role* (both positive and negative) in shaping or governing these consequences.

Engineering is widely perceived as a *profession of solving problems* and thus engineers and engineering students are proud of themselves being *problem-solvers* (Jonassen, 2014; Zhu, 2021). There are certain problems that are global in nature and transcend national boundaries, requiring collaborative efforts between engineers and non-engineers. The National Academy of Engineering (NAE) in the United States call these global concerns "Grand Challenges," highlighting the need for engineers to collaborate across borders and disciplines, drawing upon their creativity, imagination, and expertise from all engineers. However, philosopher Carl Mitcham argues that what is missing in engineering education is more than simply technical abilities to address these grand challenges (Mitcham, 2014). There is a more fundamental challenge encountered by engineering educators and engineering students, the "self-knowledge" challenge, which refers to the ability to "think reflectively and critically about the kind of world we wish to design, construct, and inhabit in and through our technologies" (Mitcham, 2014).

From Mitcham's perspective, what distinguishes between his approach and NAE's Grand Challenges is essentially the *nature and scope* of these challenges. In other words, are the problems confronted by engineering educators more of global technical problems as indicated in NAE's Grand Challenges which are expected to be solved by expertise and innovation provided by engineers? Or, do engineering educators have a more pressing moral obligation to address the challenges concerning themselves or the lack of self-knowledge in the profession? Even for highly complex questions such as Global Challenges, is engineering the only and always the best way to approach these issues? Karwat et al. (2015) argue that what is more important for addressing questions such as climate change is more of a cultural change within the engineering profession rather than engineering expertise or engineering solutions *per se*. Thus, it is crucial to educate engineering students to examine some bigger questions concerning the purposes of their work and who and whose benefits they serve (Karwat, 2019). These questions are meant to help engineers develop a self-critical consciousness and help them better understand the motivations for doing engineering work and the implications for their work. All these questions are concerned about their *moral project* not simply the engineering project. Some example questions include:

- Why am I an engineer?
- For whose benefit do I work?
- What is the full measure of my moral and social responsibility (Karwat, 2019, p. 1329)?

Readers who are familiar with engineering education research (EER) as a field in the United States may argue that most recent scholarship in EER does emphasize certain concepts that seem to have a self-dimension such as self-efficacy and self-concept (Dirisina & Shehab, 2023). It is true that these concepts do touch on how engineering students perceive or evaluate themselves. However, we would argue that discussions on these concepts miss certain engagement opportunities for students. For instance, scholarship in EER often considers the *instrumental* value of these concepts for pursuing or persisting in engineering as their career pathway. Exploring these self-related concepts in EER is mainly to understand the psychological mechanisms for engineering learning. Doing so will help to improve students' efficiency in grasping technical concepts and skills and thus help them better at solving technical problems. The *moral* value of these concepts is not addressed in these efforts. In other words, if my own values are not aligned with how I see or experience engineering, is engineering a career path I should pursue? Or if there are values or assumptions embedded in engineering practice that are questionable? If I continue to pursue engineering, what do I learn and grow as an engineer?

The discussions on these self-related concepts in EER often does not take engineering or technology seriously, including their impacts on social and human development. For instance, by participating in a particular engineering field, whose interests do I serve and what kind of society and societal values do I build? The more I participate in these engineering projects, what kind of person will I become? Is that kind of life as an engineer worth pursuing? Therefore, the term self is not merely considered as a psychological or learning science concept (in other words, understanding the self is essential for learning more efficiently). Instead, it is more of an ontological or moral concept, and it is valuable not *only* for the sake of learning successes (e.g., good self-concept skills predict academic outcomes) but also for some ontological issues such as who we are as engineers and who we want to become. Therefore, our major goal in this chapter is to leverage resources from Confucian ethics and explore how these resources can help us conceive an alternative approach to engineering education that focuses on self-cultivation.

12.3 Confucian Theory of Self-Cultivation: A Comparative Perspective

To integrate the self-critical dimension into engineering education research and practice, we aim to leverage holistic approaches to education in both the East and West traditions. More specifically, our discussions hinge on a central concept in Confucianism self-cultivation (*xiusheng*, 修身) and compare it with a similar concept *Bildung* derived from the European tradition. We expect that adopting such a comparative perspective will enhance our ability to engage in a meaningful discussion about an alternative approach to engineering education that prioritizes critical self-cultivation.

As a central concept and practice in Confucianism, self-cultivation needs to be examined within its rich historical context. As a philosophical tradition, Confucianism emerged during the Spring and Autumn Period (*chunqiu shiqi*, 春秋时期) (770 BC–476 BC). This historical period witnessed a decline in the central authority of the Zhou dynasty (1046 BC–256 BC), leading to increased political autonomy among regional lords who governed their own territories as independent states. This era witnessed the emergence and active participation of the *shi* (scholar, 士) class, or "scholar-officials" (Li, 2018), marked by a significant degree of academic freedom and diverse and yet competing intellectual traditions. During such a period marked by warfare and social upheaval, scholars gave rise to various schools of thought, seeking to provide solutions. In this era, a notable phenomenon known as the "*baijia zhengming*" (contention of hundred schools of thought, 百家争鸣) showcased academic freedom. Influential philosophers such as Laozi, Mozi, and, notably, Confucius played pivotal roles in shaping the intellectual landscape during that period.

Amidst such societal upheaval, a central question concerning all schools of thought was how society could reach social harmony. The solution that Confucius offered was through the practice of self-cultivation. Specifically, according to the Confucian perspective, a primary aim of self-cultivation is to attain the status of a *junzi* (morally exemplary person, 君子). Originally, the term *junzi* referred to a lord's son, but Confucius expanded the term to represent a "morally exemplary person." Central to this process of self-cultivation is the cultivation of virtues (or virtuous tendencies) which is expected to lead to ethical conduct, with a notable assumption that achieving this noble goal does not depend on one's social background. Therefore, ideally anyone has the potential to become a *junzi* through learning (Gardner, 2014). The ultimate aspiration becoming a *junzi* is to become a person whose behaviors in their interactions with others demonstrate the highest moral integrity, guided by cardinal virtues such as *ren* (benevolence, 仁), *yi* (righteousness, 义), *li* (ritual propriety, 礼), *zhi* (wisdom, 智), and *xin* (trustworthiness, 信). In general, self-cultivation serves as the foundation for governing larger spheres such as one's family and their state, as laid out in *Daxue* (大学).

Building on the three-dimensional model of Confucian self-cultivation proposed by Lai (2003) and the Confucian notion of self as inherently relational (Wong, 2004), we argue that self-cultivation in Confucianism covers three key elements: the self, relationships, and knowledge: (1) the process of self-cultivation involves deep self-reflection to refine one's character; (2) the practice of virtues and refinement of moral character are conducted in one's social interactions with others; and (3) lifelong pursuit of knowledge and continuous self-improvement constitute the *epistemological* basis for self-cultivation. On the basis of the three elements, the Confucian approach to self-cultivation also recognizes the evolving nature of the self, underscoring the importance of lifelong learning and self-improvement to attain the true essence of *ren* and maintain virtuous behavior throughout one's life (Zhu, 2021). For example, according to the idea of "examining oneself three times a day" (*san xing wu shen,* 三省吾身) proposed by Zengzi (曾子), one needs to examine whether they are faithful in their endeavors, trustworthy in their friendships, and committed to continuous learning on a daily basis. Further, self-cultivation is also exemplified by the concept of *neixing bujiu* ("when self-reflection reveals that nothing goes wrong, then there is nothing to be anxious about", 内省不疚), advocated by Yan Yuan (颜渊), Confucius' most favorite disciple.

Bildung, a philosophical concept derived from the Western context, has been observed to exhibit resonance with the practice of Confucian self-cultivation by scholars. *Bildung* has its European tradition and cultural roots during the Enlightenment that ushered the era of secularization in an emerging civil society (Biesta, 2002). Situated in that historical context, *Bildung* demonstrates a transformative concept of humanity with a progressive ethos: "a being who can and should take responsibility for itself and its world" (Gjesdal, 2015). As noted in Sjöström et al. (2017), two core elements are included in the concept of *Bildung*: an idealized representation of desirable knowledge and skills, coupled with processes of free and self-directed learning. It should also be noted that the concept of *Bildung* is primarily associated with intellectual and educational traditions in Germany, and still remains a contested concept ever since (or, an "empty concept", see Reichenbach (2014)).

Bildung could also be understood through its connection with the German literary genre known as *Bildungsroman* (*Bildung* + *sroman* (novel)), which delves into the mental and moral development of the protagonist, typically within the context of their coming-of-age journey. Hence, in a *Bildungsroman*, a central aim is the process in which the protagonist achieves maturity, and this transformation is usually portrayed as a challenging process that the protagonist encounters. For instance, in Johann Wolfgang von Goethe's *Wilhelm Meister's Apprenticeship*, Wilhelm Meister grapples with various conflicts, such as those between familial expectations and societal norms when embarking on different apprenticeships in fields like the theater, literature, and business, and eventually he has changed from an aimless, young dreamer to a mature, self-aware, and well-rounded adult (Moretti, 2000).

Throughout history, much like the diverse interpretations of Confucian self-cultivation, *Bildung* has been subject to varied perspectives and understandings. For instance, thinkers such as Johann Gottfried Herder, Friedrich Schiller, and Wilhelm von Humboldt considered *Bildung* as a secular phenomenon, connecting it to the emotional, moral, and intellectual growth of a person, their educational experiences and even their process of becoming a responsible citizen. For example, University of Berlin could be seen as an embodiment of the Humboldtian initiative toward a *Bildung* education during the Prussian reforms (Nybom, 2003). A Hegelian approach, in contrast, seeks to disrupt the mere dichonomy between the subjective and the objective with a dialectical approach, resulting in *Bildung* internalizing "the public or objective sphere (that is, the world in its historical development)" in "the individual, who is thus (in virtue of the dialectical method) led from subjective belief to objective knowledge or wisdom" (Nordenbo, 2002, p. 350).

Overall, *Bildung* places a strong emphasis on the significance of knowledge as a means to achieve individual freedom or a more holistic self (Gjesdal, 2015). One should also be aware that *Bildung* as an idea of general education was firstly directed mainly towards the emerging middle class, individuals who did not enjoy the privileges of the clergy and the aristocracy, with less attention given to peasants or workers. But as time went by, *Bildung* had also been applied to a broader use in the educational setting (Ringer, 1989; Løvlie & Standish, 2002). Even in a more contemporary context, where *Bildung* was misused and criticized as merely a commodity or value (Reichenbach, 2014), or where debates persist regarding the necessity of *Bildung* education (see, e.g., Masschelein & Ricken, 2003), proponents like the European Union continue to argue that the sustainability of democracy hinges on citizens with *Bildung* (Andersen, 2021), or a more proactive call for rethinking *Bildung* from a posthumanist approach (Taylor, 2017).

As mentioned above, Confucian self-cultivation and *Bildung* both aimed to help individuals navigate the shifts in their societies during certain periods of time, whether those transitions were a result of wartime disruptions or secularization. Both also underscored the importance of education for the individual. Educational scholars have recently systematically compared Confucian self-cultivation with *Bildung*. According to Yang (2022), the two concepts share at least two similarities. The following paragraphs mainly summarize Yang's (2022) work on how Confucian self-cultivation and *Bildung* share both similarities and differences in certain aspects.

First, both the two concepts emphasize the critical role of individual effort and agency in moral development. Confucian self-cultivation underscores the inherent equal potential within every individual to attain the status of a *junzi*, or morally exemplary person. The fact everyone is born with similar moral potential invites us to stay true to ourselves. Therefore, the path to becoming a *junzi* lies in actively exercising one's agency and freedom in everyday life, engaging in thoughtful observation and reflection on both others' and one's own moral choices. This introspective process should be used to inform and enrich one's future decisions. The practice of integrating reflective insights into decision-making not only aids in evaluating the plausibility of these insights but also enhances one's ability to continually observe and reflect on the moral choices made by others and themselves. Such an iterative process motivated by one's agency is central to Confucian self-cultivation. From the perspective of *Bildung*, one has the agency and freedom to train one's reasoning ability including the freedom to understand all available options and thus make informed choices based on such knowledge.

Second, both Confucian self-cultivation and *Bildung* underscore the connection between individual growth and development and the broader social and communal harmony. In Confucian ethics, we know that individual self-cultivation is fundamental for managing family (*qijia*, 齐家), governing the country (*zhiguo*, 治国), and ensuring security all under heaven (*pingtianxia*, 平天下). Similarly, in the context of *Bildung*, one aspires to become a responsible citizen, encompassing the notion that individuals, through their education and personal development, should be inclined to actively engage in public life and contribute to the welfare and progress of their community and society. Therefore, so-called "public-spirited individuals" are motivated by a sense of civic duty to promote the common good through participating in civic activities, advocating for social progress, and working towards collective well-being.

While there are similarities between Confucian self-cultivation and *Bildung*, it is also critical to recognize that these two concepts exhibit certain differences. First, they demonstrate different worldviews. Confucian self-cultivation finds its roots within the context of the *tianxia* (天下) worldview, which seeks to attain a profound state of harmony between the individual and the cosmos. This worldview is manifested on multiple levels, encompassing the self, the family, the state, and ultimately the concept of *tianxia*. In contrast, a Humboldian, tradition approach to *Bildung* revolves around the dichotomy of I and non-I, wherein the broader community, society, and even the cosmos is categorized as "non-I." The primary emphasis of *Bildung* is therefore on individual development and self-cultivation, with the societal or even cosmic connection viewed as supplementary to the overarching goal of personal growth.

Second, Confucian self-cultivation and *Bildung* adopt different approaches to conceptualizing personality and natural tendency of human beings. Confucian ethics considers that all persons are born with comparable human tendencies or moral potential for becoming good persons. Therefore, everyone is almost equal in terms of the potential to become a morally developed person or a *junzi*. In comparison, *Bildung* emphasizes the diverse attributes humans are born with. It places importance on recognizing and nurturing the varied qualities and characteristics inherent in individuals from birth. In this context, *Bildung* implies a recognition of the inherent diversity among people, encompassing a range of innate talents, qualities, and potentials that contribute to their unique identities. Such an emphasis on diversity underscores the idea that education and personal development should take into account and value the multifaceted nature of individuals from the outset of their lives.

Third, while both Confucian self-cultivation and *Bildung* underscore the role of environment in shaping one's process of becoming a good person, they diverge in their focal points. Self-cultivation puts more emphasis on a person's own determination or agency in developing and perfecting one's moral self through reflective engagements with others. However, *Bildung* highlights the value of environmental diversity, which provides diverse opportunities for individuals with distinct personalities, as central to one's moral growth.

Finally, in this section, readers may concern that our discussion and attempt to compare the two concepts may fall short in addressing broader societal or structural factors that influence individuals, thereby advocating for an individualistic moralism. In other words, when contemplating one's inner self or focusing on personality development, there is a risk of overlooking how individuals are inherently connected to a larger social structure. Nevertheless, we did mention earlier in this section, both the two traditions do emphasize the connection between individual growth and development and the broader communal harmony and social order. Moreover, at least in the Confucian tradition, the virtue of *yi* (义, appropriateness) is comparable to the Western concept of justice at least in the Aristotelian sense (Sim, 2010).

Acting in accordance with the virtue of *yi* can ensure the alignment of personal conduct with broader social moral principles and help cultivate the moral sensitivity among people toward the underlying social and structural factors that affect personal growth.

In the next section, we will synthetize the similarities and differences between Confucian self-cultivation and *Bildung* and utilize this comparative perspective to formulate an approach to engineering education grounded in self-cultivation.

12.4 Toward a Self-Cultivation Approach to Engineering Education

Based on a comparative review of the Confucian concept of self-cultivation, this section explores how such a concept can provide implications for constructing an alternative approach to engineering education that focuses on the self-dimension of the professional formation of engineers.

A first step toward such a goal is to critically examine the learning outcomes often found prevalent in engineering education, including how they are often formulated, what they often entail, and how they influence curriculum priorities, assessment, and pedagogy. Learning outcomes are arguably the most fundamental component in the American approach to engineering education. Historically, a critical move to turn engineering education into a field informed by educational sciences and evidence was to adopt the outcome-based or backward design approach (Wiggins & McTighe, 2005). Such a philosophy has been integrated into both accreditation policies and classroom practices.

More specifically, effective design of learning activities in engineering starts with a *pragmatic* question concerning learning outcomes: what do we expect students to achieve after they complete a specific learning activity? Given that engineering education is informed by the outcome-based, backward design approach, what engineering educators are supposed to care about would only be to what extent students achieve these learning outcomes after they participate in a particular learning activity. It does not matter much what instructional methods are employed to achieve these outcomes insofar as they achieve these outcomes. It provides freedom to both instructors and students and allow them to achieve these goals through diverse ways. This raises the crucial question: how can we *empirically* verify that students have indeed attained these outcomes? Engineering educators thus need to ensure that good learning outcomes are supposed to be supported by observable, empirical (either quantitative or qualitative) evidence of students' learning achievements (Davis & Beyerlein, 2006). Therefore, good learning outcomes thus typically manifest as either knowledge or skills. Nevertheless, it is worth noting that it is not easy to collect empirical assessment data for all learning outcomes.

Consider, for instance, the widely recognized engineering learning outcomes, such as the ABET student learning outcomes. The most recent version of student learning outcomes for accrediting engineering programs include:

1. an ability to identify, formulate, and solve complex engineering problems by applying principles of engineering, science, and mathematics;
2. an ability to apply engineering design to produce solutions that meet specified needs with consideration of public health, safety, and welfare, as well as global, cultural, social, environmental, and economic factors;
3. an ability to communicate effectively with a range of audiences;
4. an ability to recognize ethical and professional responsibilities in engineering situations and make informed judgments, which must consider the impact of engineering solutions in global, economic, environmental, and societal contexts;
5. an ability to function effectively on a team whose members together provide leadership, create a collaborative and inclusive environment, establish goals, plan tasks, and meet objectives;
6. an ability to develop and conduct appropriate experimentation, analyze and interpret data, and use engineering judgment to draw conclusions;
7. an ability to acquire and apply new knowledge as needed, using appropriate learning strategies (ABET, 2023).

It is evident that most these learning outcomes are formulated based on either technical knowledge (e.g., principles of engineering, science, and mathematics) or skills (e.g., solutions that meet specified needs with consideration of public health, safety, and welfare, as well as global, cultural, social, environmental, and economic factors) (even ethics is considered as an instrumental skill for engineering design). These learning outcomes largely assume that developing these practical abilities is central to the formation of competent engineers and it is critical for "solving complex engineering problems" (as seen in Objective 1) and transforming the society. As previously noted, gathering empirical assessment evidence for all learning outcomes can be challenging or impractical. This challenge typically leads to two outcomes: First, certain learning objectives, such as ethics, may be simplified, reducing them to technical skills, such as selecting the "best" solution from multiple predetermined options. Second, some learning outcomes may be discarded altogether due to their difficulty to measure accurately. Hence, exisiting engineering accreditation criteria may overlook certain essential outcomes.

For instance, what is missing in the learning outcomes in existing engineering accreditation criteria is to what extent students can develop an ability to critically examine the *broader impacts of engineering work on themselves*. Instead of solely emphasizing how engineering expertise can solve technical problems and transform the world, can we envision an alternative learning outcome centered on students' capacity to evaluate *how their engagement in engineering shapes and is shaped by their self-knowledge, values, and beliefs*?

Considering the pivotal role of reflective learning in both Confucian self-cultivation and *Bildung*, an additional learning outcome might be articulated as follows: Engineering students demonstrate the ability to critically observe the practices of participants in the engineering community, including not only their peers but also various stakeholders such as managers, investors, and workers. They adeptly integrate insights acquired from reflective observations into their own work, and eventually lead to continuous improvement of themselves. In other words, assessing the quality of engineering education hinges on the extent to which students are capable of competently developing technical solutions but also on their capacity to cultivate an awareness of continuous self-improvement through reflective learning in professional settings.

From a more holistic perspective, another challenge with most learning outcomes in engineering today is that they do not explicitly address the fact that engineering students may come to engineering with diverse backgrounds. Confucianism does emphasize the argument that every person is born with the same potential for development especially moral growth. However, when students start their undergraduate engineering programs, there have already been variations among them due to different learning experiences they have each encountered since they were born. In comparison, *Bildung* emphasizes the diverse attributes humans are born with. It places importance on recognizing and nurturing the varied qualities and characteristics inherent in individuals from birth. *Bildung* implies a recognition of the inherent diversity among people, encompassing a range of innate talents, qualities, and potentials that contribute to their unique identities. Such an emphasis on diversity underscores the idea that education and personal development should take into account and value the multifaceted nature of individuals. When engineering educators formulate and implement learning outcomes, they need to keep in mind that students all come with different qualities and therefore these learning outcomes may mean differently to them. An overarching objective for engineering educators involves effectively engaging with individual students, adeptly tailoring learning outcomes to suit diverse situations. This approach aims to enhance learning experiences by aligning them more closely with the distinctive needs, values, and career aspirations of each student.

Finally, engineering education extends beyond the realm of individual professional and moral growth. Both Confucian self-cultivation and *Bildung* advocate for the connection between the growth of individual engineers and the broader social harmony or progress. A major limitation with the current form of engineering learning outcomes is that these outcomes only focus on *individual development within engineering*. Engineering educators should strive to educate their students to critically examine how their expertise can be socially relevant by serving the needs of people in various communities. Engineering students are thus invited to critically reflect on how their expertise is (or should be) used in the workplace, whose interests their work serve, what values their work contributes to, and if and how these values help promote the wellbeing of the public. The lead author of this chapter used to employ various engaged pedagogies such as ethics autobiographies and personal ethics statements to cultivate students' capacity to reflect on the relationship between their daily, "mundane decisions" and the implications for broader social structures (Snieder & Zhu, 2020; Zhu & Woodson, 2020).

Both Confucian self-cultivation and *Bildung* acknowledge that various human (personal) and organizational resources present in the environment can facilitate the self-improvement of engineers. Earlier in this section we already articulated the value of engagement with others in professional settings for engineers' self-improvement. Moreover, certain forms of organizational or structural resources can also be helpful for engineers' growth. For instance, choosing a company that values responsible innovation and corporate social responsibility (CSR) is beneficial for engineers' cultivation of moral character (Smith et al., 2021). Therefore, educating engineering students to be cognizant of the ethical dimension of career choices is critical for engineering educators. In addition, engineering students need to be aware that different values and moral tendencies can be potentially cultivated while they are assigned to work with individuals and organizations from diverse backgrounds. Collaborating with individuals solely focused on the profitability of their products may cultivate different values and moral tendencies compared to working with people from underserved communities, where basic human needs are barely met.

While the concept of self-cultivation offers valuable insights for engineering education, it is important to recognize that it may present challenges and limitations, particularly when applied across diverse cultural contexts. To effectively implement this approach in engineering education, at least two key factors are essential. First, there must be a *philosophical foundation* that supports efforts to cultivate the moral self of students, especially considering that Western engineering education tends to focus heavily on individualistic, employable skills. Second, even with a philosophical foundation for the self-cultivation approach, there needs to be an *infrastructure* within engineering education to translate this mindset into concrete practices. This includes classroom pedagogy, university policies, curriculum design, and student support programs. Unfortunately, changes in these two factors require large-scale institutional transformations, which can be particularly challenging in Western contexts due to the democratic governance structures prevalent in higher education. These structures often involve extensive consultation and consensus-building processes among diverse stakeholders, including faculty, administrators, and students, which can slow down the process of implementing substantial changes in curriculum and instructional practices.

In summary, this chapter does not aim to formulate a comprehensive framework for engineering education centered on the concept of self-cultivation. Instead, it should be viewed as an initial endeavor to investigate the potential utility of the idea of self-cultivation in crafting an alternative approach to prevalent methods in engineering education, which predominantly concentrate on the development of *employable* skills for individual engineers. Future research is needed to discuss how the idea of self-cultivation can help design engaged learning activities and assessment strategies that promote self-cultivation, continuous self-improvement, and ultimately a flourishing life for engineers.

Acknowledgments This material is based on work supported by the National Science Foundation under Grant No. 2124984. Any opinions, findings, conclusions, or recommendations expressed herein are those of the author(s) and do not necessarily reflect the views of the National Science Foundation. Additionally, this work was partially funded by the Program for Leadership and Character at Wake Forest University and Lilly Endowment Inc.

References

ABET. (2023). *2023–2024 criteria for accrediting engineering programs*. ABET. Available at: https://www.abet.org/2024-2025_eac_criteria/

Andersen, L. R. (2021). *What is Bildung? And how does it relate to ALE?* European Association for the Education of Adults (EAEA). Available at https://veranstaltungen.vhs.or.at/programm/bereichsuebergreifende-angebote/what-is-bildung-a-new-bildung-agenda-for-europe/

Biesta, G. (2002), How general can Bildung be? Reflections on the future of a modern educational ideal *Journal of Philosophy of Education, 36*(3), 377–390.

Davis, D., & Beyerlein, S. (2006). Deriving design course learning outcomes from a professional profile. *International Journal of Engineering Education, 22*(3), 439–446.

Dirisina, J., & Shehab, R. L. (2023). *A systematic review of academic self-concept measures in first-year engineering education*. American Society for Engineering Education Annual Conference, Baltimore, MD.

Downey, G. L. (2012). The local engineer: Normative holism in engineering formation. In S. H. Christensen, C. Mitcham, B. Li, & Y. An (Eds.), *Engineering, development and philosophy* (Vol. 11, pp. 233–251). Springer.

Gardner, D. K. (2014). *Confucianism: A very short introduction*. Oxford University Press.

Gjesdal, K. (2015). *Bildung*. In M. N. Forster & K. Gjesdal (Eds.), *The Oxford handbook of German philosophy in the nineteenth century*. Oxford University Press.

Herkert, J. R. (2000). Engineering ethics education in the USA: Content, pedagogy and curriculum. *European Journal of Engineering Education, 25*(4), 303–313.

Jonassen, D. H. (2014). Engineers as problem solvers. In A. Johri & B. M. Olds (Eds.), *Cambridge handbook of engineering education research* (pp. 103–118). Cambridge University Press.

Karwat, D. M. A. (2019). Engineering for the people: Putting peace, social justice, and environmental protection at the heart of all engineering. In *National Academy of engineering. Frontiers of engineering: Reports on leading-edge engineering from the 2018 symposium*. National Academies Press. https://www.ncbi.nlm.nih.gov/books/NBK538716/

Karwat, D. M. A., Eagle, W. E., Wooldridge, M. S., & Princen, T. E. (2015). Activist engineering: Changing engineering practice by deploying praxis. *Science and Engineering Ethics, 21*(1), 227–239.

Lai, K. (2003). Confucian moral cultivation: Some parallels with musical training. In K.-C. Chong, S.-H. Tan, & C. L. Ten (Eds.), *The moral circle and the self: Chinese and Western approaches* (pp. 1070–1139). Open Court.

Li, S. (2018). *The constitution of ancient China*. Princeton University Press.

Løvlie, L., & Standish, P. (2002). Introduction: *Bildung* and the idea of a liberal education. *Journal of Philosophy of Education, 36*(3), 317–340.

Masschelein, J., & Ricken, N. (2003). Do we (still) need the concept of *Bildung*? *Educational Philosophy and Theory, 35*(2), 139–154.

Mitcham, C. (2014). The true grand challenge for engineering: Self-knowledge. *Issues in Science and Technology, 31*(1), 19.

Moretti, F. (2000). *The way of the world: The Bildungsroman in European culture*. Verso.

Nordenbo, S. E. (2002). *Bildung* and the thinking of *Bildung*. *Journal of Philosophy of Education, 36*(3), 341–352.

Nybom, T. (2003). The Humboldt legacy: Reflections on the past, present, and future of the European university. *Higher Education Policy, 16*(2), 141–159.

Reichenbach, R. (2014). "Humanistic *Bildung*": Regulative idea or empty concept? *Asia Pacific Education Review, 15*(1), 65–70.

Ringer, F. (1989). *Bildung*: The social and ideological context of the German historical tradition. *History of European Ideas, 10*(2), 193–202.

Sim, M. (2010). Rethinking virtue ethics and social justice with Aristotle and Confucius. *Asian Philosophy, 20*(2), 195–213.

Sjöström, J., Frerichs, N., Zuin, V., & Eiks, I. (2017). Use of the concept of *Bildung* in the international science education literature, its potential, and implications for teaching and learning. *Studies in Science Education, 53*(2), 165–192.

Smith, N., Zhu, Q., Smith, J., & Mitcham, C. (2021). Enhancing engineering ethics: Role ethics and corporate social responsibility. *Science and Engineering Ethics, 27*(3), Article 28. https://doi.org/10.1007/s11948-021-00289-7

Snieder, R., & Zhu, Q. (2020). Connecting to the heart: Teaching value-based professional ethics. *Science and Engineering Ethics, 26*(4), 2235–2254.

Taylor, C. A. (2017). Is a posthumanist *Bildung* possible? Reclaiming the promise of *Bildung* for contemporary higher education. *Higher Education, 74*, 419–435.

Wiggins, G. P., & McTighe, J. (2005). *Understanding by design* (2nd ed.). Pearson.

Wong, D. (2004). Relational and autonomous selves. *Journal of Chinese Philosophy, 31*(4), 419–432.

Yang, L. (2022). Student formation in higher education: A comparison and combination of Confucian xiushen (self-cultivation) and *Bildung*. *Higher Education, 83*(5), 1163–1180.

Zhu, Q. (2021). Self-cultivation of the Confucian engineer: What engineering ethics education can learn from Confucian moral theory. In X. Wang & P. Wong (Eds.), *Harmonious technology: A Confucian ethics of technology* (pp. 66–78). Routledge.

Zhu, Q., & Woodson, S. (2020). Educating self-reflective engineers: Ethics autobiography as a tool for moral pedagogy in engineering. *Teaching Ethics, 20*(1/2), 31–46.

Qin Zhu Associate professor of engineering education at Virginia Tech. He is an associate editor for *Science and Engineering Ethics* and an Executive Committee member of the Society for Ethics across the curriculum (SEAC). His major research interests include the cultural foundations of engineering ethics, global engineering education, and the ethics and policy of robotics and computing technologies. His most recent work examines the ethical values communicated in the "hidden curriculum" in engineering and the design of robots that contribute to a flourishing "moral ecology" of human-robot interaction.

Yixiang Shawn Sun Ph.D. student in Engineering Education at Virginia Tech. He is also an affiliate non-resident Junior Fellow at Research Institute for Democracy, Society, and Emerging Technology, Taiwan. He holds a BA in Sociology and an MS in Geography from National Taiwan University. His research interests include engineering/AI ethics education, global engineering education, comparative education, and global talent migration. Overall, he aims to situate his research within the present-day geopolitical context, with Taiwan as a starting point and a comparative approach as a guiding methodology.

Chapter 13
Humanities and Engineering Education for Gaston Berger from the Perspective of *Bildung*

Marie-Pierre Escudié

Abstract This chapter proposes to re-read Gaston Berger's philosophical work on education from the perspective of *Bildung*. Although the concept does not appear directly in his work, it is possible to identify several theoretical and practical influences related to this concept that are present in his writings and actions. Several dimensions of Berger's thought take on new meaning in relation to education, such as the need to think in terms of humanity, to confront disciplines in order to rethink knowledge, and to integrate the goal of the social world into education. In order to reflect on these educational axes, it is appropriate to return to the influence of Edmund Husserl on Gaston Berger. In this chapter, we will try to make a rapprochement between Berger's theoretical attitude, inscribed in his affinity with the German phenomenologist, and *Bildung*.

With John Dewey, Berger makes a pragmatist turn in the sense that he also makes experience the motor of humanistic education. The empirical and experimental educational principles that Berger shares with Dewey can be analyzed in terms of their contribution to *Bildung*. On the one hand, the spirit of *Bildung* sheds an original light on Gaston Berger's thought and can be found subtly in the new institutions of French higher education after 1950; on the other hand, it is not so much a question of looking for the scientific presence of *Bildung* in Berger as of making this past a factor of appreciation of the present in relation to the new doctrines of ecological transition (social responsibility and sustainability).

Keywords Gaston Berger · Engineering school · INSA · Education · Theory · Foresight · Ethics · Humanism · Technology

---

M.-P. Escudié (✉)
INSA Lyon, Institut Gaston Berger/Centre des Humanités, Villeurbanne, France
e-mail: marie-pierre.escudie@insa-lyon.fr

A. Buch, S. H. Christensen (eds.), *Bildung for Engineering Education and Practice*, Philosophy of Engineering and Technology 49,
https://doi.org/10.1007/978-3-031-86581-7_13

13.1 Introduction

The reinterpretation of Gaston Berger's educational philosophy in terms of *Bildung* invites us to go in search of the embodiment of an idea and to investigate a concrete utopia.

Philosopher, entrepreneur, teacher at the University of Aix-Marseille, combatant in the First World War and Resistance fighter in the Second World War, Director General of Higher Education from 1953 to 1960, Gaston Berger's active life symbolizes his philosophy. In France, he is not considered a major philosopher of the twentieth century, but he was one of the first to introduce Husserlian phenomenology and a pioneer of foresight.

His writings on education make no reference to *Bildung*, but I assume that they are largely imbued with this vision. The presence of this thought extends even beyond his texts, collected in a French edition under the title L'homme moderne et son éducation (Berger, 1962), to feed his work as a senior administrator in higher education and the institutional creations that followed.

From a French perspective, the philosophy of *Bildung* cultivates self-improvement, the perfection of talents for the individual and the community, "throughout life and in existential forms that are not necessarily institutionalized" (Fabre, 2019). Despite the general theoretical distancing of French thought from *Bildung*, the intention of this chapter is to examine the singular effort to institutionalize *Bildung* as intended and implemented by Gaston Berger.

The creation of the French engineering school Institut National des Sciences Appli-quées (INSA) in Lyon in 1957 is a case in point. INSA is philosophy in action (Escudié, 2013). In the manner of Wilhelm von Humboldt, Gaston Berger sought to apply his philosophical principles on education to educational institutions. The creation of the Institut National des Sciences Appliquées in Lyon in 1957 was inspired by the model of the Technische Hochschulen and the Massachusetts Institute of Technology. In studying the INSA, we will focus on the creation of a humanities department that symbolizes the relationship between "humanism and technology," bringing an original contextual renewal to education.

This chapter presents Berger's main pedagogical principles in constant relation to *Bildung*, and then goes on to analyze the philosophical background of his work to put *Bildung* in context. The third part describes Berger's process of conversion, which adds a spiritual dimension to his educational thinking. The pragmatic dimension of Berger's work concludes the study, linking thought to action.

13.2 Politics of the Spirit of the Humanities: Some Educational Principles of *Bildung*

In order to question the connections between Berger's work and *Bildung*, I propose to explain his intention in terms of a politics of the spirit of the humanities. On the one hand, there is a reference to Paul Valéry's "politics of the spirit" (Valéry, 1953), crossed, on the other, with what can be described more broadly as the "spirit of the humanities" in the philosophy of education (Nussbaum, 2010).

What was the background to the implementation of the politics of the spirit of the humanities? In the mid-twentieth century, the understanding of an accelerating world, on a human scale, renewed the purpose of education, which was too tradition oriented. Berger draws on the poet and epistemologist Paul Valéry's idea of the need to think about the politics of the spirit. Education and the future are intimately linked and determine our concept of humanity. Education has the anthropological function of understanding and making people. Berger advocates education and teaching that prepares the individual to "confront what has never been"[1] (Berger, 1962, p.129). So, first and foremost, we need a clear vision of future humanity, its role, its aspirations, and its possibilities. According to this point of view, the sciences can say things about the world, but they can only be appreciated in terms of the human intelligence projecting itself into the future. In contrast to *Bildung*, the "politics of the spirit" is concerned with education in an idealistic rather than a practical sense, in which truth meets the virtue of education. To carry out this project, Berger relies on the humanities, whose vocation is to form critical consciousness and sensitize the soul. Contemporary philosopher Martha Nussbaum describes the "spirit of the humanities" as the pursuit of critical thinking, bold imagination, empathic understanding of human experience in all its diversity, and an appreciation of the complexity of the world in which we live (Nussbaum, 2010). This politics of the spirit of the humanities is exemplified by the creation of the Humanities Department at INSA. Berger emphasized the development of imagination, creativity, and critical thinking, drawing on the heritage of the humanities disciplines: rhetoric, history, ethics, and poetry.

Berger's humanistic education is expressed in several practical dimensions, based on a variety of classical and European philosophical origins that are compatible with the classical vision of *Bildung*.

[1] My translation. All quotes are from my own translation.

13.2.1 Academic Freedom and Autonomy for Educational Institutions

Berger's political action was based on a philosophical stance. As a professor himself, he saw the mission of the university in terms of freedom of action and institutional independence. Berger warns that "Free of his actions, master of his work, sole judge of what he must undertake, the professor must never forget that he works for the community of men and that he must, consequently, situate himself in this community." (Berger, 1961, p.71–72). This attitude of educators is closely related to the role of the philosopher in society. From a Platonic perspective, the philosopher (or educator) must have the freedom to choose the problems to be addressed. His task is to restore humanity's sense of purpose and wholeness. It is not directly the management of human affairs or the government of the state that the philosopher-educator brings to bear, but the enlightenment of the context and values of the time for government.

There are two basic educational principles in Berger's work. The first is that teaching, which imparts knowledge, must give way to education, which forms people. Berger explains that the two purposes of school are to give children the knowledge they will certainly need and to prepare them to become citizens. "If we confine ourselves," he says, "to the first point, to things which are really useful, which must be perfectly known, but which are relatively few in number, there will be plenty of time for the thorough and leisurely pursuit of one or two studies which will give the students a real culture." (Berger, 1962, p.135).

Although Berger's project refers to the Renaissance model of studia humanitatis, it is by no means a humanism of erudition and encyclopedism. The rediscovery of ancient texts, which gave access to the heritage of past civilizations, opened the way to an in-depth, comparative study of historical, cultural, and religious values. Through a process of metamorphosis, however, this culture reaches the contemporary age with the aim of transforming it, not freezing it. For Berger, the humanistic aspects of each science—and of the social sciences as well, as we will see later with prospective anthropology—must be restored to the center of education, so that knowledge can take on meaning and serve to orient oneself in the world. This form of intelligibility is something that "all teaching can provide, if it is presented in a certain spirit" (Berger, 1962, p.135).

The second principle is programmatic: *education must be lifelong*. In a society characterized by movement and the precariousness of values, it is becoming impossible to rely solely on traditional knowledge. Education is no longer a question of content, but of method. It must be seen as a "willingness" to learn and think continuously, at a level of abstraction that prevents us from falling into the apathy of mechanical (intellectual) gestures. Berger's ambition for education in general is to open "the book of the world" (Méry, 1964) to active learning outside of libraries and universities.

With respect to these principles, the example of INSA shows that the implementation of a propaedeutic year as an integral part of the training is designed to provide a solid educational foundation while stimulating curiosity about the sciences and the humanities. However, this is not unique to INSA, as the classical preparatory class for Grandes Ecoles in France favors this model of general theoretical knowledge. Disciplinary orientation and vocational training should not come too early, since it is intellectual training that guarantees later access to and interest in lifelong learning. This vision of knowledge is reminiscent of the liberal arts model in the United States according to Berger, which precedes vocational orientation.

13.2.2 A Social Perspective on Bildung

In addition to the intellectual dominance of the model of *Bildung*, it is interesting to look at the social dimension that makes this model possible. In its ideal of education for emancipation, *Bildung* integrates the material conditions necessary for the realization of a secure life. These are not really principles of social justice for equality and solidarity, but a moral concern for the good life, provided by appropriately thought-out practical means. The question of housing is invested in the construction of university campuses and is also found in the working-class models of the "familistères" (a place where several families or individuals live together in a kind of community. In addition to housing, several buildings such as schools, cultural and social facilities contribute to the common good). In contrast to the Humboldtian model, English education included a social dimension by providing housing for their students. This principle, which developed in Germany at the beginning of the twentieth century, was initially incorporated into the INSA project. The school functions like a campus, with private living spaces—the so-called "thurnes"—and public spaces—the restaurant, library, and performance venues.

The social aspect developed elsewhere in Europe or in the American campus model both distances and revises the aristocratic form of German *Bildung*. The appearance of the INSA in the landscape of French civil engineering schools was precisely intended to break with the elitism of the traditional engineering schools, such as the Ecole Polytechnique, founded in 1795 or even the Ponts et Chaussées, founded as early as 1747. This criticism of elitist culture and class distinction is reflected in the figures quoted in the press at the opening of INSA: "20.45% of the INSA students come from working class and peasant families" (Butheau, 1962/07.07).

At the crossroads of intellectual and social aspects, there is no competitive entrance examination at INSA. Contrary to the classical model, the selection is made without long prepared tests, in favor of a free conversation with a multidisciplinary jury (teachers, psychologists and engineers). In its first years, the school organized a winter sports discovery week, given Lyon's proximity to the Alps, during which evenings were devoted to lectures by professionals from the world of entertainment, medicine and so on. As part of the curriculum, an internship is compulsory at the end of the first year in order to put the future engineer in the situation of the employees he or she will have to manage. The winter sports week was abandoned, mainly due to the sharp change in the social background of the students, who were already familiar with this type of leisure activity. Other policies changed rapidly as INSA evolved, for example, in the early years it was possible to enter the school without a baccalaureate, due to the many school careers interrupted by the war years of the previous decade, but this was soon discontinued. One of the original aims of INSA was to promote a greater diversity of social backgrounds among its engineering students by decentralizing the school to the provinces.

13.2.3 Unity in Teaching, Research and Society

Berger explains that research and teaching must be considered together, with substantial time allocated to research. Research requires human, financial, and technical resources to foster the true "inventors" that the post-war world was then demanding. But Berger goes further, demanding that research be shared, communicated with fervor to a wider audience than just the scientific community. In his view, thinking about a sustainable problem does not mean isolating oneself completely in the laboratory, but exposing oneself to the critical judgment of a demanding public in order to make progress. Berger even sees it as a duty to transfer knowledge from science to society. We are witnessing the beginnings of action research and the debate between the sciences and social actors, in the sense that Berger will even create an institution for this purpose, the Centre International de Prospective, in 1957.

When it comes to research, Berger supports the idea of not separating science from industry. Industry is about concrete experience. It links science to situations in which knowledge is synthetically mobilized. Berger affirms the primary character of the experience from which science is "used". Science tackles a problem in order to find an answer. It serves to analyze, quantify, and investigate a question drawn from the concrete world, using intellectual and scientific means. And in this perspective, the confrontation of several disciplines allows "mutual enlightenment" (Berger, 1961, p.77). Industry, as the locus of concrete problems and effective action, is both the point of departure and the point of direction for the sciences.

Berger distinguishes between basic and applied research, but does not oppose them. The meaning of what he calls in French "sciences utilisées" situates the context as the object of science. Science is projected into the world, learning from situations as well as explaining them.

13.2.4 Culture as a Sense of the Human

In keeping with *Bildung*, Berger asserts that "the university is not the culmination of studies, but an opening to life" (Berger, 1962, p.74). By refusing to confine itself to a narrow specialization that is incomprehensible to most people, university education seeks to communicate to society the "human significance of what it deals with" (Berger, 1962, p.72). In this way, I'll use the image that the purpose of education is to keep the fire of culture burning, which for Berger means putting the meaning of the human being back at the center of thought and action.

The purpose of liberal arts education is to restore culture and bear witness to humanity through the creative imagination. Disciplines such as the arts, literature, languages, and civilization offer a plurality of human meanings. They finely describe concrete situations through a variety of words and images, as well as cultural and spiritual traditions.

In addition to the idealistic pursuit of culture, Berger emphasized the need to study concrete people, characters, and actions. To this end, he created the Humanities Department in the 1950s. Around the existing humanities department, "through him and under him" (Guyon, 1964, p.7), humanistic education was expanded and epistemologically renewed. Under the influence of the humanities, the word "culture" was transformed from the scholarship of the past to the consciousness of modern society. Berger explains that "knowledge of the environment in which [man] operates, from which he receives incentives and where he develops his initiatives, is all the more necessary because the concrete man is always bound to a certain situation," engaged" in a certain context" (Berger, 1962, p.152).

Berger wants to link the question of technique with culture in order to qualify this context. Technique and culture are not antithetical. Technique, as a human endeavor elaborated by intelligence, can heuristically nourish culture. Indeed, technology provides new evidence for human action, and in turn, culture can provide a sense of direction for action when technology, while increasing action tenfold, also creates anxiety. For effective action in the world, Berger calls for the humanities to use technology to bring them closer to the concrete world. This idea has implications for education, insofar as it resituates cultural ambitions in technical vocational training. From this perspective, Berger defends the need to think about technology in and through culture.

The Humboldtian model is opposed to the proponents of vocational training. It is based on the idea that one must first and foremost be an upright person and a good citizen. On this basis, professional skills are easily acquired over time, in line with scientific, technological and administrative developments. This model of education seeks to avoid abandoning the education of well-rounded citizens capable of thinking for themselves in favor of technical qualifications and skills geared to global market competitiveness (Nussbaum, 2010). Berger's combination of a professional school for engineers and a liberal arts education is consistent with this vision.

13.3 The Background to the Crisis of Intelligibility

For Berger, the meaning of the human adventure is the conquest of knowledge. Conquest symbolizes both problem and test. As Jennifer A. Herdt writes, "we awoke to discover humanity as task" (Herdt, 2019, p.1), and so as we study the world, we must study ourselves. Berger's work is relentlessly concerned with the connection between the conditions of knowledge and the subject who thinks them. For Berger, philosophy is not so much knowledge as an attitude and available attention.

While rigorously pursuing the intelligibility of the mind, he believes, like Paul Valéry, that it is impossible to abandon the problem of life altogether in order to deal exclusively with the mind. Thus, the goal of this work remains concrete: to solve human problems through a philosophy of mind. There is unity of consciousness only when words and deeds are united.

Since the sources of experience lie outside the subject, an inner attitude is essential. Berger takes up the Husserlian method of epoché, which consists in freeing oneself from the world (moving from nature to the freedom of the mind) and then engaging the world with a new judgment. Epoché is an exercise in combating the prejudices inherent in our character and our environment. In the twentieth century, this philosophical exercise was to be profoundly affected by the historical events of war, the Holocaust, and the atomic bomb, which forced us to start from scratch.

13.3.1 The Influence of Crisis: The Crisis of the Intelligibility of the Spirit

The influence of the crisis of the early twentieth century is at the center of Berger's thinking. Quoting extensively from Husserl and Valéry, he uses the concept of "crisis" to describe Europe's mental disorder as the starting point for his philosophical action on behalf of culture.

Three years before Edmund Husserl's lecture to the Vienna Cultural Circle in 1935, Paul Valéry's Critique also denounced this spiritual disorder and raised the major consequence of man's inability to imagine the future. This preoccupation with the future is of interest to Berger in that he himself finds it almost entirely absent from philosophical thought. Valéry points to the scientific-technical paradox of an "im-masterable mastery" (Valéry, 1936) that loses sight of both the purpose of knowledge and the appropriateness of its applications in the world.

The crisis of the mind arises from a situation in which philosophy, whose task it is to describe the content of the mind and its operations, no longer satisfies the need to know itself (Le Senne, 1970). This is a mission that must be constantly undertaken, because the concept of the spirit does not refer to a closed world; the spirit is concerned only with a subjective experience that cannot fully contain it. But spirit is always outside reality and can no longer be grasped or oriented. Its aim is to promote the development of consciousness, restoring to the spirit its critical role, as well as "the effort towards wisdom, as an overcoming of the partial, egocentric, selfish self, to reach the level of a higher self that sees all things, in the perspective of universality and totality, that becomes aware of itself as part of the cosmos, which then embraces the totality of things" (Hadot, 2001, p.142).

Berger links the crisis in the intelligibility of the world to a loss of spirituality. Without determining whether this is a cause or a consequence, the meaning of the world has gradually been imposed by science and engineering. The assertion of philosophy's role in the face of this is to connect the intelligibility of the world and human action to spiritual or ethical ends in order to restore a broader meaning to experience. In this perspective, Berger insists that engineering, whatever its field of application, must not simply become a technique that triumphs over reflection (Berger, 1962). Against technical solutionism or blissful optimism in progress, Berger will address two critiques of modern science.

13.3.2 Criticism of Naturalism and Scientism

It is possible to draw a parallel between the Humboldtian model and Husserl's famous lecture on the crisis of the spiritual discourse that underpins European humanity. While the Humboldtian model is opposed to the natural and social sciences, Husserl defends the idea that the sciences, inspired by naturalism, have distanced themselves from the ambient world of life. This leads to a modern dualism in the interpretation of the world that separates man from nature. More precisely, this dualism claims that the unconditional truth produced by science is distinct from the relative truths of the real world. Thus, it is the analytical dimension of science that determines experience, in such a way that phenomena are divided, decomposed, and then reunited.

To combat this crisis of humanity, whose foundations of scientific ideality have forgotten the relationship between science and the concrete world, we need to rebuild a "science of the whole world" (Husserl, 1976). By this, Husserl means the rediscovery of cultural unity on the basis of the system developed by the Greeks: philosophy. The goal is to teleologically reaffirm the historical-philosophical idea of European humanity.

The natural sciences have developed a relationship with the closed world, understood through a method of pure mathematics and objective physics, the exact elaboration of which, according to Husserl, has largely enabled "the technical domination of nature" (Husserl, 1976, p. 349). The sciences of the mind, going back to ancient Greece, have represented the world subjectively, giving value to the "ambient world," both the totality and the individualities that carry infinity within themselves. Husserl's lecture warns against the "misunderstanding" that would claim to give the sciences of the mind the support of a science of nature. Conversely, he emphasizes the folly of extracting the sciences of nature from the services of the mind through naturalistic blindness or an excess of modern dualism.

On the contrary, we need to reunite scientific and extra-scientific thinking. This opens up the possibility of rediscovering the way of being of humanity, which is no longer based on everyday empiricism, but on objective truth. The German philosopher thus presupposes a science of the mind with a scientific project capable of exercising scientific and critical self-knowledge. In other words, in his view, intentionality must be placed back at the center of the task of knowledge. For Husserl, the historical outcome of this crisis lies in the alternative between European decline driven by spiritual hatred and European rebirth based on philosophical heroism.

When Berger takes up this critique of naturalism, he exposes the problem of never questioning the being under consideration, of making it an absolute fact of nature, which makes it possible to make a moral judgment based on the workings and dysfunctions of that nature. Reducing knowledge to the matter and form of the object reintroduces one of the great dualisms of the modern distinction between nature and culture. "Knowledge escapes us, he insists, as soon as we try to enclose it in a nature, even if we super-impose on phenomenal nature a hypothetical and unverifiable transcendent nature that is only a faded copy of the former." (Berger, 1941, p. 31).

Berger formulates a critique of positivism and scientism as a religion of man and progress. In the blind and deaf search for the objectivity of fact, subjectivity is reduced to nothing in this way of thinking. The ambition is such that.

> [...] the interposition of recording devices between the observer and the thing studied often makes it possible to eliminate errors due to fatigue, carelessness, distraction, malice, and so on. But the reduction to a simple reading has not eliminated personal examination. It has undoubtedly been distanced, diluted, and minimized. We have also - and this is what we wanted - made it anonymous. (Berger, 1941, p. 28).

Positivism lacks the "presence" of the thing in consciousness, that is, the operation of linking fact and statement. This critique justifies recourse to philosophy in the sense that, for Berger, it restores a sense of purpose and totality.

13.4 The Act of Conversion: The Spiritual Dimension of *Bildung*

To overcome the criticism of the scientific method, Gaston Berger, still under the influence of Husserl, developed the project of a theoretic.

Theoretics provides a philosophical starting point for examining the testing of knowledge by consciousness. As a science of the mind, theoretics consists in adopting an attitude that distinguishes between the I, as subject, and the individual caught up in his or her worldly experiences. The goal is to give humanity its true meaning through the contribution of all the sciences, rethought according to a new intelligibility. The starting point is Berger's critique of modes of knowledge that lack consciousness. On this point, I hypothesize that *Bildung* has a homology with this reflection on the modalities of knowledge at the service of consciousness. Theoretics can then nourish the philosophical and spiritual foundations of *Bildung*.

13.4.1 *Theoretics: The Personal Test of Knowledge*

The following paragraph outlines the stages in Berger's development of the theoretical attitude:

> There is no valid philosophy without respect for objectivity, without an interest in science and, consequently, in the modern world, without an interest in technology, which confirms the solidity of science and develops it into applications that it is up to us to make useful. Nor is philosophy without a fundamental need for intelligibility. The philosopher serves the truth, and technical success or the strength of feeling are, at their best, only presumptions of truth. They cannot be neglected, nor can we be satisfied with them. There is no philosophy without metaphysics. Without metaphysics, religion is no more than a social structure, and science a collection of more or less effective recipes. The law of the object makes sense only to a subject - to a subject and not to a man. (Berger, 1962, p. 29).

The question of objectivity is not worked out separately from subjectivity in Berger, and this includes the relationship to scientific knowledge: "The circle is never fully recognized until it is traced by us as recreated by our own act." (Berger, 1941, p. 52). At this stage, the subject is not outside of perception, any more than the center of the circle is inside the circle. Only limited, imperfect knowledge is possible, for it derives its meaning precisely from the paradox that "there is no vision except through opacity" (Berger, 1941, p. 100). In the case of a scientific experience, as in the case of a personal ordeal, knowledge can only be fixed by a certain language or gesture. These languages are light and shadow, they say or don't say, demonstrate or don't demonstrate. As a reader of Paul Valéry, Berger discusses the pencil stroke, "[…] that smear on the white page - which allows forms to appear that can then be charged with meaning" (Berger, 1941, p. 100), expressing his discovery of Leonardo da Vinci's sfumato, which Valéry also analyzed as such a subtle form of knowledge (Le Moigne, 2007).

Knowledge through experience shows that there are intentions of meaning, whether theoretical, moral, or emotional. Objectivity is not indifferent to values. Existence is always grasped in a certain way, by a center of preference, because perception expresses a possible way of grasping knowledge. "Because it is situated, says Berger, my knowledge is limited and imperfect" (Berger, 1941, p. 98). In order to achieve the level of intelligibility he envisions, Berger embraces the Husserlian concept of the suspension of judgment. This involves the practice of systematic doubt in order to achieve progressive detachment. This detachment invites us to set aside theories, systems, preconceptions, symbolic operations, and representations.

To reach the cogito, we must free ourselves from the grip of society and the pressures of our psychophysiological ego. Detachment from the cogito is neither indifference nor objectification. It is impossible to suppress one's own situation, but it is advisable not to give it more importance than the situation of others. Detachment seeks to know the world, without bringing the meaning it offers back to oneself. Detachment does not give access to an inner fullness, but rather to an existential barrenness—no ego-sufficiency—that requires an "inner demand". The individual who becomes a subject "[…] perceives himself as an infinite striving to which corresponds a call from a total transcendence" (Berger, 1962, p. 319).

In Berger's Descartes-inspired metaphysical perspective, God is the creator of values and essences and the guarantor of the rationality of the universe. In the Cartesian approach, if the statement is based on a metaphysics present outside the subject, this means that scientific discourse is immediately given the foundations of all its own validity and objectivity. For Berger, the sciences must lead to metaphysics through an inner quest that remains hidden a priori. Scientific research alone does not allow for detachment from the world; it does not contain the truth, but is a "presumption of truth. By reversing the place of metaphysics as he does, Berger criticizes and limits a form of rationalism that places the reality of God outside the world. Human action, whether scientific or cultural, cannot have absolute values if it is linked to a primary metaphysics; on the contrary, it tends toward the absolute only through an intentional act of research.

13.4.2 Faust's Parable: Linking Theoretics and **Bildung**

The regaining of intelligibility through the conversion of the subject refers to a transcendental and committed meaning. It is impossible to get out of the world, and it is only with the intention that it is possible to escape subjectivity. The transcendental makes possible an understanding of the world that is grasped in its existential unity, in its relation to value, and in intersubjective knowledge. Berger chooses to illustrate this approach with a important reference from *Bildung*.

At a conference for Lafarge factory workers in 1960, Berger personally reinterpreted the end of Faust. He reads Goethe's work as follows: Dr. Faust is bound by a pact with Mephistopheles to attain absolute knowledge. A continental traveler in search of truth, Faust succeeds in freeing himself from the devil when he lands in a remote village in the Netherlands, where he is to make healthy and fertile swampland unfit for agriculture. By serving the community, he makes sense of the encyclopedic knowledge he has acquired and can consciously free himself from the sealed pact. Berger's view of this work focuses on the intentionality of the engineer Faust, who finds the humanist principle of humility at the center of his actions and then puts it into practice in engineering. Berger explains:

> [...] the technician Faust of old is a Faust who, with modesty, has accepted to play his role as a man. He pays the price. He knows that in this limited, imperfect world, in this world of time that is ours, every result is the fruit of labor. It's not enough to say that every result is the fruit of intelligent work. And that's what technique is all about. Technology is both the application of intelligence to work and the provision of the fruits of labor to mankind. Without human generosity, technology is meaningless. Without the rational application of intelligent processes, technology is ineffective. It is for this reason that the myth of Faust is, I believe, singularly instructive (Berger, 1962, p.141).

A theoretical attitude serves a genuine engagement with the world. It opens the door to understanding and acting in the world:

> It's not me I'm grasping, it's 'the other', it's the world, in which I see myself as existing neither more nor less than the rest, a small piece of the world alongside the others, joyful

> moreover at this rediscovered familiarity with all other men, who are, in the full force of the term, my fellow men; with other living beings; with things too, which we are wont to call inanimate. They are caught up in the same web as I am, and their existence prolongs mine (Berger, 1941, p. 47).

Berger, who freshly and powerfully celebrates the ways of being of the living and the non-living, does not seek to isolate consciousness. Intelligibility becomes interdependent with the intimacy of each being: "I resume my place in the great family of beings; I rediscover the comradeship of the animal, the more discreet friendship of the tree, the fidelity of the earth" (Berger, 1941, p. 47). I interpret this form of intelligibility in an ecological sense, as a science of relationships, but without naming it as such. Berger links the human and the metaphysical adventure in the cosmos. Berger's transcendental idealism is even more demanding of the transcendental subject, in that it is impossible to obtain completely certain knowledge. And this shortcoming, admitted but not overcome, demands a pragmatic and responsible ethics of knowledge.

13.5 The Practical Reason

In practice, *Bildung* is fundamentally ethical and political. On the one hand, its educational mission is to educate autonomous individuals capable of self-determination and responsibility through reason. On the other hand, it aims to create citizens of the world who are equally capable of addressing the major issues facing humanity (not job-oriented).

The search for this educational ideal was a necessity in the context of the beginning of the second half of the twentieth century. Berger's inspiration for *Bildung* is in fact a way of reconnecting with values, reactivating them, appropriating them, and refining them in a direction that orders certain values with others. His work brings together the dimensions of the individual and the citizen under the term "man". The interest of this term lies in the connection between the human essence and the concrete human condition, in which we must situate ourselves.

13.5.1 Links with Pragmatism

Berger's eulogy at the time of Dewey's death makes it interesting to examine Berger's proximity to the pragmatist influence. On the question of education, Berger's critique of traditional education is shared. Indeed, both share a vision of a journey from school to life. This leads their vision of the educational project towards a democratic goal, comparable to the ideal of education (Sjöström et al., 2017).

Berger's interest in the pragmatist method stems from his recognition of the retreat of "system philosophies" and the heuristic relevance of the concepts of truth and certainty. As a lesson from the traumas of his century, " we distinguish between what seems certain and what we think we understand. Philosophy becomes a gradual investigation instead of a picture painted once and for all, all at once" (Berger, 1962, pp. 279–280). Berger thus emphasizes the need for a pragmatic philosophy that questions situations in relation to a "totality that serves as a context" (Berger, 1962, p. 56) in a universe in constant reconstruction.

In the articles he wrote on foresight in the 1950s, Berger focused on developing the sciences of action, in which calculating the consequences of our actions implies an ethic of responsibility. The pragmatic dimension is more about resolving conflict, as Berger says in his tribute to John Dewey, than about freeing our minds. This development transforms our relationship with knowledge, which loses its static character. The inertia inherent in its transmission must evolve, according to Berger. Education, as mentioned above, must become continuous education in order to respond to the problems of the real world. Dewey's deconstruction of the assumption that the future will resemble the past, suggesting that change will become the rule, is largely echoed by Berger.

This is reflected in his reflections on action-oriented professions, including engineering, where Berger rejects the "theatrical choice" between the philosopher detached from the world and the man of action dominating the world. Neither total efficiency nor absolute purity, the true human task is to seek "valid efficiency" for a given value horizon. Action in the world requires a dynamic relationship with knowledge, a need for synthesis, a "synoptic" vision, and the emergence of new knowledge, which Berger seeks to address through a prospective anthropology.

13.5.2 Ingenium

Without mentioning Giambattista Vico, Berger reflects on ingenium. In his research on the faculties of knowledge, he points out that "[…] ideas of synthesis or analysis are not pure intellectual data. Their immediate appeal lies in the sensitive quality they embody. Linking ideas, synthesizing them, analyzing difficulties, is always a bit like picking up or separating with my hands" (Berger, 1941, p. 10). By paying attention to the embodied subject, Berger's vision of the ingenium is nourished by a phenomenology of the body-subject in relation to the world around it, which fundamentally connects them. A philosophy of the ingenium, but also a science of the ingenium, insofar as Berger values recursive reasoning (Le Moigne, 2007), which cannot, however, certify its scientific discourse with the seal of truth: "It is impossible to distinguish here [for the practical sciences], as is done without too much difficulty in other disciplines, between direct action and reflected knowledge: every action is reflexive and every reflection is creative" (Berger et al., 2008, p. 40). The synthetic endeavor transforms the relationship to truth through a pragmatic process of verification.

As a reflective attitude, ingenium can also be understood as an attitude of creativity. And it is precisely to the figure of the engineer that Berger addresses the theoretical attitude with a practical concern. "The engineer, he says, does not use in his daily practice any truth that does not come from a particular science, but it is up to him to make an original synthesis of them" (Berger, 1962, p. 141).

13.6 Conclusion

References to *Bildung* are formally absent from Berger's work, but borrowings and interpretations are quite explicit in his actions. In his biography, Berger also embodies the spirit of *Bildung*. "It's not just a theory I'm expounding, it's an experience I'm summarizing." (Berger, 1962, p.145).

I must now consider the contribution of this reflection to the contemporary context of the implementation of a policy of sustainable development and social responsibility in academic training. In the dominant epistemological configuration of an engineering school, the question of intelligibility refers to the status of knowledge, its ontologies, and methods, in a negative relationship with subjectivity. Too much empirical subjectivism would run the risk of allowing philosophy, or even politics, to invade all scientific fields and subject knowledge to opinions or relative truths. Returning to the spirit of *Bildung* and Berger's foundational theoretical approach means rethinking the role of subjectivity and exploring other meanings of knowledge. In addition to this philosophical intention, new epistemologies drawn from the humanities concretize and pluralize this relationship to knowledge, for example through theories of the situated point of view, or the recognition of experiential, local, and lay knowledge, or the place of prospective imaginaries. This last point, which has its origins in Berger's thought, is one of the ways in which the ecological and social transition developed by INSA in recent years can restore critical autonomy to students and citizens.

In particular, this raises an important debate, linked to the heritage of *Bildung*, on humanism and ecology. In recent years, INSA Lyon has been implementing a wide-ranging training reform, which will be voted on by the school's board of administration in 2019. Its objective is to deploy 24 ECTS over the five years of sustainability training, in line with UNESCO's international guidelines in 2017 (UNESCO, 2017) and national with Jean Jouzel's Report in 2022 for higher education. Its common core curriculum includes the main themes of the Anthropocene and planetary limits, climate, energy, living issues, resources, levers for action, as well as possible and desirable futures... This evolution in curricula not only has the consequence of redefining the way we envisage the roles and responsibilities of engineers in society, but also of rethinking contemporary humanist values. The challenge before us is to reconsider humanism in the Anthropocene in the face of a conservative humanism (Morizot, 2018) based on several dimensions in movement: rethinking a relational and not ontologically substantial and essentialist humanism, situated between particularistic needs and a universalist ideal, situated also from vulnerabilities and beyond modern anthropocentrism to re-envision—modestly?—human perfectibility in connection with the Earth. The historical and philosophical depth of *Bildung* is a critical resource to ponder, to give due consideration to inter-relational roles in their relationship to the world, when they sometimes seem to falter in the face of the scale of the task. When confronted with the challenges of environmental and social justice, this notion enables us to reinvest the values of respect, peace and solidarity in the exercise of critical reason and sensibility in the face of contemporary global upheavals. Because it contains an ethical and political requirement, the current context imperatively invites us to take up the *Bildung* project of training a citizen engineer capable of assuming his or her own responsibilities through a close and intimate anthropological knowledge of living beings.

References

Berger, G. (1941). *Recherches sur les conditions de la connaissance*. PUF.

Berger, G. (1961). *Prospective*, 7.

Berger, G. (1962). *L'homme moderne et son éducation*. PUF.

Berger, G., de Bourbon-Busser, J., & Massé, P. (2008). *De la prospective. Textes fondamentaux de la prospective française 1955–1966*. L'Harmattan.

Butheau, R. (1962-7-7). « 20,45% des étudiants de l'INSA sont issus de familles ouvrières et pay sannes ». Le Progrès.

Escudié, M.-P. (2013). *Gaston Berger, les sciences humaines et les sciences pour l'ingénieur. Un projet de réforme de la société*. Université de Lyon.

Fabre Michel, « Bildung », in Christine Delory-Momberger éd., Vocabulaire des histoires de vie et de la recherche biographique. Toulouse, Érès, « Questions de société », 2019;197–199. https://doi.org/10.3917/eres.delor.2019.01.0197. https://www.cairn.info/--9782749265018-page-197.htm

Guyon, B. (1964). *Hommage à Gaston Berger. Actes du colloque du 17 février 1962 à Aix-en-Provence*. Publication des Annales de la Faculté des Lettres d'Aix-en-Provence.

Hadot, P. (2001). *La Philosophie comme manière de vivre*. Albin Michel.

Herdt, J.-A. (2019). *Forming humanity. Redeeming the German Bildung tradition*. The University of Chicago Press.

Husserl, E. (1976). *La crise des sciences européennes et la phénoménologie transcendantale*. Gallimard.

Jouzel, J. (2022). Sensibiliser et former aux enjeux de la transition écologique et du développement durable dans l'enseignement supérieur. URL: https://www.enseignementsup-recherche.gouv.fr/fr/sensibiliser-et-former-aux-enjeux-de-la-transition-ecologique-dans-l-enseignement-superieur-83888.

Le Moigne, J.-L. (2007, février). Les enjeux éthiques de la recherche et de l'intervention en éducation et formation n'appellent-ils pas un « nouveau discours de la méthode des études de notre temps » ? *Réseau Intelligence de la complexité*. URL: https://www.intelligence-complexite.org/media/document/conseil_scient/intelligence-complexite-enjeux-ethiques-recherche-et-lintervention/open

Le Senne, R. (1970). *Introduction à la philosophie*. PUF. (1939).

Méry, M. (1964). *Gaston Berger et le « livre du monde»* (Hommage à Gaston Berger) (Vol. 42). Editions Ophrys.

Morizot, B. (2018). L'écologie contre l'Humanisme. Sur l'insistance d'un faux problème. Essais, 13. URL: http://journals.openedition.org/essais/516

Nussbaum, M. (2010). *Not for profit – Why democracy needs the humanities*. Princeton University Press.

Sjöström, J., Frerichs, N., Zuin, V.-G., & Eilks, I. (2017). Use of the concept of *Bildung* in the international science education literature, its potential, and implications for teaching and learning. *Studies in Science Education, 53*(2), 165–192. https://doi.org/10.1080/03057267.2017.1384649

UNESCO. (2017). *Education for sustainable development goals: Learning objectives*. URL: https://www.unesco.org/en/articles/education-sustainable-development-goals-learning-objectives
Valéry, P. (1953). *La politique de l'esprit. Variété III*. Gallimard (1936).

Marie-Pierre Escudié A teacher at INSA Lyon (Institut National des Sciences Appliquées), she holds a PhD in political science entitled *Gaston Berger, les sciences humaines et les sciences de l'ingénieur. Un projet de réforme de la société* (2013). A member of the Institut Gaston Berger, her research focuses on engineering ethics and the social responsibility of engineers. She teaches modules in human and social sciences, ethics, and ecological transition at INSA's Centre des Humanités.

Chapter 14
Conclusion: *Bildung* for Engineering Futures

Anders Buch

Abstract This chapter synthesizes the anthology's contributions to explore how the concept of *Bildung* becomes relevant as a response to evolving challenge perceptions in engineering education and practice. The authors of the various chapters rejuvenate the concept of *Bildung* in different ways, each shaped by their distinct perceptions of contemporary challenges. Drawing from diverse intellectual traditions, they highlight different aspects of the complex, situated sociotechnical challenges facing engineering today. This underscores that *Bildung* is a normative endeavor that can no longer assume a stable and independent environment. As *Bildung* enters the Anthropocene era, articulating its vision for engineering must involve an active reconfiguration of social norms and practices to foster good and just ways of life.

Keywords *Bildung* · Engineering · Futures · Normativity · Challenge perception · Anthropocene

14.1 Introduction

The concept of *Bildung* emerged from Enlightenment visions in Western culture and was articulated by the philosophers, educators, and poets of Romanticism in late eighteenth-century Europe. It was preceded by ideas of *Paideia* in ancient Greece (cf. Buch & Christensen, 2025) and similar concepts in other ancient cultures (cf. Zhu & Sun, 2025). Since Romanticism, ideas of *Bildung* have evolved and transformed in various historical and cultural contexts. However, the concept has primarily been associated with general education, upbringing, and acquaintance

A. Buch (✉)
Via University College, Valby, Denmark
e-mail: buch@via.dk

A. Buch, S. H. Christensen (eds.), *Bildung for Engineering Education and Practice*, Philosophy of Engineering and Technology 49,
https://doi.org/10.1007/978-3-031-86581-7_14

with the fine arts, and it has rarely been considered as a foundation for addressing higher education or specialized professional activities.

Drawing inspiration from Sjöström and his colleagues (Sjöström et al., 2017; Sjöström & Eilks, 2018) and their discussion of *Bildung* in contemporary STEM (Science, Technology, Engineering, and Mathematics), the authors of this anthology have explored the viability of these ideas in modern engineering education and practice. Sjöström et al.'s mapping of the different pathways through which the notion of *Bildung* has traveled has been instrumental in exploring the potential of broadening STEM education to include social and cultural elements. Their work has shown that the concept of *Bildung* has resonated with various sentiments in different cultural contexts over two centuries, and the authors of this anthology attest that its relevance endures in engineering education and practice today. The enduring agenda of *Bildung* aligns with contemporary considerations about the future direction of engineering education and professionalism.

But why does this old agenda resurface in contemporary visions of STEM education and engineering professionalism? And how should it be adapted to address modern concerns? In the introduction (Christensen & Buch, 2025), we noted that the notion of *Bildung* generally tends to reemerge in response to crises—when challenges cannot be addressed within existing practices. When the current educational system and professional institutions can no longer fulfill their promises to provide the necessary competencies and skills to tackle present and future challenges, educators, politicians, and industry leaders begin to reconsider the options for adjustments or even major changes in education and professionalism. Sociotechnical imaginaries of desirable futures often frame discussions about education and professionalism, suggesting strategies for addressing challenges and achieving objectives (Jasanoff & Kim, 2015; Buch et al., 2024). However, the processes of identifying, classifying, and labeling pertinent challenges, and envisioning effective strategies to solve them, are interrelated. In envisioning desirable futures, values and objectives are mobilized that implicitly define the challenges that must be overcome to realize the desired future state of affairs—and identifying challenges presupposes the existence of desirable futures that are currently obstructed.

In engineering, perceptions of challenges and response strategies are coupled in various ways in competing visions of the profession's desirable futures. Some proponents emphasize the proficiency and effectiveness of engineering in optimizing economic growth as the primary objective, while others advocate for a broader professional and technological outlook concerning societal responsibility. Still, others envision the goal of engineering as transcending current notions of technological proficiency and effectiveness to include more holistic sociotechnical competencies needed for sustainable transitions in society (cf. Buch, 2012 for a governmentality analysis of contemporary challenge perceptions and response strategies in engineering). The tropes of the *Bildung* agenda are often invoked by proponents who envision futures where engineers play a central role in fostering just and sustainable transitions in society, and who, accordingly, define the challenges in these terms—both on a personal level, through self-cultivation, and collectively, through joint and engaged action.

14.2 Challenge Perceptions

Considering the notion of *Bildung* in engineering education and practice, the authors of this anthology identify challenges to be overcome and strategies to be pursued for the engineering profession. Anders Buch and Steen Hyldgaard Christensen (2025) argue for a *Bildung* perspective in engineering that begins with the democratic, cultural, and human challenges of society, reconstructing the vision of *Bildung* by drawing on Dewey's democratic philosophy. Carl Mitcham warns that unreflective trust in the promises of technology is itself a challenge that engineers must recognize as a *sine qua non* of responsible professionalism. Diana Adela Martin et al. (2025) propose Challenge Based Learning as a contemporary relational pedagogical principle to reinvigorate *Bildung*, countering sociotechnical challenges identified by stakeholders in local ecosystems. Renaud Prouteau and co-authors (2025) suggest a reciprocal articulation of *Bildung* and citizenship through a paradigm of existential citizenship, calling for political action and ecological awakening to overcome challenges. Eddie Conlon, from a Marxist perspective, reconsiders *Bildung* (2025) as a transformative force to humanize the world and help emancipate humanity from the ravages of capitalism by promoting deep and active learning in engineering programs. Conversely, Karl Palmås (2025) warns against politicizing engineering education, arguing that the classical *Bildung* objective of enabling students to "think for themselves" remains crucial to avoid standard critiques of technology and present the world to students in an unbiased manner.

Appelhans and Cheville (2025), along with Kjelsberg and Kahrs (2025), take an empirical approach to studying *Bildung*. Appelhans and Cheville explore *Bildung* by examining the teaching of engineering from a contextual and societal perspective at a U.S. liberal arts university. By situating a course within the liberal arts tradition of *Bildung*, they document the challenges of teaching the course and students' responses. Similarly, Kjelsberg and Kahrs analyze students' motivations for studying *Bildung*-oriented engineering and discuss the relationship between societal usefulness and *Bildung*. These two empirical chapters concretely scrutinize how different *Bildung*-oriented strategies broaden engineering education to ensure their relevance and usefulness in addressing societal challenges. James Trevelyan (2025) advocates for integrating *Bildung* into engineering education, emphasizing the importance of understanding the social and organizational contexts in which engineers work. He identifies the issue of relevant contextualization of engineering practices as a key challenge in reproducing engineering professionalism within educational institutions.

Zhu and Sun (2025) build on Mitcham's diagnosis that the real challenge to engineering education and professionalism is self-knowledge—i.e., "the ability to think reflectively and critically about the kind of world we wish to design, construct, and inhabit in and through our technologies." To address this challenge, they draw on Confucian ethics to envision an alternative approach to engineering education that emphasizes self-cultivation. Finally, Marie-Pierre Escurdié interprets *Bildung* through the lens of Gaston Berger's philosophy, suggesting this interpretation as a resource for better addressing ecological transitions and providing academic training that ensures social responsibility and environmental sustainability in engineering.

In discussing *Bildung*, many authors refer to the traditional tropes that suffused the concept in the thinking of Enlightenment thinkers and Romanticists of late eighteenth-century Europe: viewing *Bildung* as a means to overcome challenges such as increasing alienation from nature, fragmentation of the self, loss of meaning, spirituality, and individuality, moral decay, cultural impoverishment, inadequate education, etc.—and shaping the future of engineering education and practice to tackle these pertinent challenges. However, the authors also explore new horizons. Compared to its origins in late eighteenth-century Europe, their discussions of *Bildung* have absorbed the experiences of living, working, and learning in contemporary industrialized, capitalist, globalized, and economically, technologically, and culturally differentiated (and even polarized) societies. Enlightenment's visions for society have not been fully realized, and the Romanticists' skepticism has been validated. Confronting well-established higher education systems, an established engineering profession, and a world saturated with technological solutions, the authors observe that old challenges persist, and new challenges have emerged. Carl Mitcham (2025) articulates these new challenges most succinctly: "The classical theory and practice of *Bildung* focused on the person and assumed a stable world to which *Bildung* pursues a relationship. What does it become when the world becomes unstable? What does it become for those engineers who are intentional or unintentional contributors to world instability?"

14.3 *Bildung* in the Anthropocene Era

The new challenges are characterized by the fact that we can no longer presume that our strategies respond to a stable and independent world. Human activities are fundamentally changing and shaping the world we inhabit. The very idea of *Bildung*—i.e., the holistic development and cultivation of individual potentials, collective action, and societal contexts—aims, as a strategy, to transform the world. Engineers change the physical world. Industrialization, urbanization, deforestation, and pollution have dramatically altered the physical environment to serve human purposes—successfully in some respects, catastrophically in others. Likewise, ideas of Enlightenment and education have shaped our ways of interacting with one another and our relationship with the material world. In modernity, our social imaginaries have been altered to place the human individual center stage in ethical and political discourse (Taylor, 1989). We now live in an ecological mutation of nature of an unprecedented scale (Latour, 2017) and have entered what has been labeled the Anthropocene era.

In the face of this new and volatile condition, contemporary visions of *Bildung* are beginning to change. Ellen Krogh and co-authors reflect on the challenges for education in light of the concept of *Bildung* in the Anthropocene (Krogh et al., 2023). They argue that education must be rethought "...to confront and introduce global challenges, such as climate change and the Anthropocene, and societal and political questions of post-truth claims and fake news" (p. 173). Philosophers of technology have stressed that humanist *social imaginaries* about desirable present and future orders are better understood as *sociotechnical* imaginaries, i.e., "…collectively held, institutionally stabilized, and publicly performed visions of desirable futures, animated by shared understandings of forms of social life and social order attainable through, and supportive of, advances in science and technology" (Jasanoff & Kim, 2015, 4). In the Anthropocene, *Bildung* can no longer be properly understood solely in terms of Enlightenment's visions of purifying reason or Romanticist conceptions of stimulating intellectual imagination. In the Anthropocene, problems and challenges are framed as sociotechnical in nature, and the strategies of *Bildung* must become hybrid and open to the complexities and dynamics of the multitude of sociotechnical challenges encountered. The strategies for developing engineering education and practice considered by the authors in this anthology testify that conceptions of *Bildung* must be rearticulated in terms that appreciate the sociotechnical nature of the challenges of the Anthropocene. *Bildung* for engineering education and practice must rest on a sociotechnical perspective that accommodates contemporary concerns and equips engineering professionals with the knowledge and skills necessary to tackle the challenges and pursue the strategies of sociotechnical imaginaries about desirable futures. *Bildung* for engineering education and practice thus embarks on a highly normative, ethical, and political agenda guided by our values and ideas about what should be considered good and just ways of life.

14.4 Normativities in *Bildung* for Engineering Education and Practice

The role of normativities in engineering education and practice is central to the discussions of *Bildung*. In engineering, as in society at large, social activity and action are guided by actors' ends, goals, interests, and preferences. The social practices that structure these activities are informed by a telos. However, the telos of engineering practices is not static or definitively established—it evolves over time, and different actors within engineering often disagree on the ultimate goals of engineering activities.

Normativity is the key mechanism through which social order in engineering education and professionalism is constructed and through which activities are organized. The organization of engineering practices does not arise merely from shared values, tacit knowledge, or uniform rules that professional engineers are expected to follow. Instead, order emerges through the dynamic interplay of reciprocal adjustments, corrections, and sanctions among actors, driven by their stakes in the interactions. The mechanism also requires recognizing who sanctions actions, as sanctions are not only performative but also perspectival, with some actors' sanctions carrying more weight than others.

Thus, the normativities in engineering are dynamic and temporal, always directed toward future activities and concerned with how social order in the profession should evolve. Neophytes do not merely come to understand or participate in engineering practices, and old-timers do not simply perpetuate established practices; they actively shape, enact, and reimagine these practices as desired structures, continuously articulating and practicing engineering. This ongoing normative engagement, which entangles cognition and emotion, means that organizing in the engineering profession involves more than the acquisition and recollection of knowledge about rules and values; it is an active reconfiguration of social norms and practices. Recognizing this role of normativities in engineering diminishes the traditional divide between 'is' and 'ought' as actors align their practices with both current norms and future aspirations, thereby contributing to the evolving standards that guide the continuation or alteration of engineering practices. Change and continuity in engineering professionalism coexist in incremental enactments of social practices (Schatzki, 2019; Stjerne et al., 2023). In this way, organizing and social order in engineering professionalism are accomplishments based on the historical preconditions for engineers' activities and their situated enactments of engineering practices in the present, embedded in a complex interplay of innovation, commitment to tradition, and personal and collective aspirations toward achieving future ends. Joseph Rouse makes this point:

"A thoroughly processual conception of a practice-differentiated way of life recognizes practices as temporally extended patterns of situated, interdependent performance. What each participant does depends not only on past performances but also on the anticipated future continuation of the practice and others on which it depends. [...] people participate in practices for the sake of what the practice is doing, what it can or will do, and how their participation affects or shapes their own developing lives. Participants' subsequent performances, even in aiming to continue the practice and fulfill its animating concerns, depend on others' future participation and changes or continuities in the associated circumstances" (Rouse, 2023, 142).

Articulating *Bildung* for engineering education and practice is a performative discursive step toward envisioning changes in the engineering profession and imagining new ways of bringing technological expertise to serve our purposes. The final word has not been said in this anthology—nor will it ever be elsewhere. Discussions about the future role of engineering will continue. Hopefully, the perspectives on *Bildung* introduced in this anthology will help enrich those future discussions.

References

Appelhans, S., & Cheville, A. (2025). Appropriateness and value: Cultivating *Bildung* mindsets amongst undergraduate engineers in a third-year design course. In A. Buch & S. H. Christensen (Eds.), *Bildung in engineering education and practice: A new agenda*. Springer.

Buch, A. (2012). Governing engineering. In S. H. Christensen, C. Mitcham, B. Li, & Y. An (Eds.), *Engineering, development and philosophy: American, Chinese and European perspectives* (pp. 169–182). Springer.

Buch, A., Lindberg, Y., & Cerratto-Pargman, T. (Eds.). (2024). *Framing futures in postdigital education: Critical concepts for data-driven practices*. Springer.

Christensen, S. H., & Buch, A. (2025). Introduction: Contesting narrow conceptions of engineering education and practice through the concept of Bildung. In A. Buch & S. H. Christensen (Eds.), *Bildung for engineering education and practice. A new agenda*. Springer.

Conlon, E. (2025). *Bildung*: Conformity or transformation? In A. Buch & S. H. Christensen (Eds.), *Bildung in engineering education and practice: A new agenda*. Springer.

Jasanoff, S., & Kim, S. H. (Eds.). (2015). *Dreamscapes of modernity: Sociotechnical imaginaries and the fabrication of power*. Chicago University Press.

Kjelsberg, R., & Kahrs, M. S. (2025). Usefulness and *Bildung*: Engaging students in a *Bildung*-oriented engineering education. In A. Buch & S. H. Christensen (Eds.), *Bildung in engineering education and practice: A new agenda*. Springer.

Krogh, E., Qvortrup, A., & Graf, S. T. (Eds.). (2023). *Bildung, knowledge, and global challenges in education: Didaktik and curriculum in the Anthropocene era*. Routledge.

Latour, B. (2017). *Facing Gaia: Eight lectures on the new climatic regime*. Polity Press.

Martin, D. A., Spahn, A., & Bombaerts, G. (2025). From the individual to the collective dimension of *Bildung* in engineering education through challenge-based learning. In A. Buch & S. H. Christensen (Eds.), *Bildung in engineering education and practice: A new agenda*. Springer.

Mitcham, C. (2025). *Bildung* under conditions of engineered destruction. In A. Buch & S. H. Christensen (Eds.), *Bildung in engineering education and practice: A new agenda*. Springer.

Palmås, K. (2025). *Bildung*, and how that concept sits with traditional notions of teaching students "how to think" like engineers. In A. Buch & S. H. Christensen (Eds.), *Bildung in engineering education and practice: A new agenda*. Springer.

Prouteau, F., Wallenhorst, N., & Hétier, R. (2025). Thinking about engineering: Between *Bildung* and citizenship. In A. Buch & S. H. Christensen (Eds.), *Bildung in engineering education and practice: A new agenda*. Springer.

Rouse, J. (2023). *Social practices as biological niche construction*. Chicago University Press.

Schatzki, T. R. (2019). *Social change in a material world*. Routledge.

Sjöström, J., & Eilks, I. (2018). Reconsidering different visions of scientific literacy and science education based on the concept of *Bildung*. In Y. J. Dori, Z. R. Mevarech, & D. R. Baker (Eds.), *Cognition, metacognition, and culture in STEM education: Learning, teaching and assessment* (Vol. 24). Springer International Publishing. https://doi.org/10.1007/978-3-319-66659-4

Sjöström, J., Frerichs, N., Zuin, V.-G., & Eilks, I. (2017). Use of the concept of *Bildung* in the international science education literature, its potential, and implications for teaching and learning. *Studies in Science Education, 53*(2), 165–192. https://doi.org/10.1080/03057267.2017.1384649

Stjerne, I., Buch, A., & Wenzel, M. (2023). Imaginary practices as the nexus between continuity and disruptive change. In T. Hernes & M. Feuls (Eds.), *Towards a research agenda for organizational continuity and change* (pp. 127–146). Edward Elgar Publishers.

Taylor, C. (1989). *Sources of the self: The making of modern identity*. Harvard University Press.

Trevelyan, J. P. (2025). *Bildung* and education for engineering practice. In A. Buch & S. H. Christensen (Eds.), *Bildung in engineering education and practice: A new agenda*. Springer.

Zhu, Q., & Sun, Y. S. (2025). Toward a self-cultivation approach to engineering education: A Confucian perspective. In A. Buch & S. H. Christensen (Eds.), *Bildung in engineering education and practice: A new agenda*. Springer.

Anders Buch Docent at VIA University College, Denmark and affiliated researcher at Jönköping University, Sweden. He holds an MA in philosophy from University of Copenhagen and a BEd and PhD in education from Roskilde University. His empirical research area is focused on professions, professionalism, and technological expert cultures, and his theoretical approach is primarily inspired by Science & Technology Studies, Practice Theory, and classical American Pragmatism. He has published articles and books on knowledge, learning, education, professionalism, and the professional development of engineers. Since 2016 he has served as chief editor of Nordic Journal for Working Life Studies.

The manufacturer's authorised representative in the EU is Springer Nature Customer Service Centre GmbH, Europaplatz 3, 69115 Heidelberg, Germany. If you have any concerns regarding our products, please contact ProductSafety@springernature.com

Printed and bound by CPI Group (UK) Ltd, Croydon, CR0 4YY
30/09/2026
02225237-0005